TRENDS IN PRACTICAL COLLOID SCIENCE

Trends in Practical Colloid Science

Vuk Uskoković

Nova Science Publishers, Inc.
New York

For permission to use material from this book please contact us:
Telephone 631-231-7269; Fax 631-231-8175
Web Site: http://www.novapublishers.com

Library of Congress Cataloging-in-Publication Data

Uskokovic, Vuk.
Trends in practical colloid science / Vuk Uskokovic.
p. cm.
Includes index.
ISBN 978-1-60692-946-9 (softcover)
1. Colloids. I. Title.
QD549.U835 2009
541'.345--dc22 2009004758

Published by Nova Science Publishers, Inc. ✦ New York

CONTENTS

PREFACE

Practical scope of modern-day colloid science may be presented through a set of achievements and actual aims in preparation of novel materials and construction of novel technologies based on the foundation of knowledge on colloid systems. Yet many uncertainties pervade such field, ranging from the true potential of self-organizing and self-assembling systems in the design of advanced material structures to numerous encounters over the question of practical viability between the tendencies to invest in knowledge on small-scale, molecular recognition processes and on spontaneous, large-scale mesoscopic formation phenomena. Albeit the existence of belief in advancement of knowledge on stereoscopic molecular recognition and molecular assembly manipulation that would eventually lead to perfect control inherent in the design of macroscopic structures, trial-and-error phenomena seem to permeate all relevant levels of organization within practical colloid science approaches, from enzymatic, biomolecular recognition processes to the macroscopic design of novel functional outcomes. As genetic evolution teaches us, the major point of any design conductance is not elimination of inherent mistakes, but their productive acceptance, that is mutual coupling with development of improved and more richly organized contexts of knowledge. Within this perspective, inter-disciplinarity, that is constructive crossing of separate scientific areas of investigation, presents a necessary approach immanent in advancement of practical colloid science achievements in the coming era.

"Nature to be commanded must be obeyed"

Francis Bacon, New Organon, 1620

Chapter 1

INTRODUCTION

What is the real prospect of perpetual human aspirations towards perfect design, i.e. perfect copying of the products of our conceptual imagination on the substrate of reality? This is the question that this review article will aim to answer. There has certainly been a significant advance in this sense throughout the previous millennia of human history. After pictograms, geoglyphs and simple construction schemes in the past, enormously complex structures, from ultrasensitive composite bridges to tiny electronic devices, can be prepared today basing on exact procedures and theoretical models with relatively high confidence in attaining satisfying reproducibility.

With informational enrichment of the realm of human experience, the tendencies to discover novel boundaries at ever smaller sizes naturally follow. Numerous novel scientific fields have been developed throughout the history of science with the instigation of such innovative interests towards smaller. Quantum mechanics, molecular biology and from recently nanosciences and nanotechnologies present some of the examples from the previous century. However, as the scientists come up with discoveries that belong to ever smaller sizes, it is also getting ever harder to use such data for a practical design of advanced functional structures and devices. Complex pathways of microscopic world are being ever more difficult to relate to regularly manipulated experimental variables of macroscopic world. In order to obtain fascinating structures that will functionalize some of the 'small' effects, it becomes necessary to rely on the design framework of colloid science. Semiconductor quantum dots, mesoscopic molecular patterns or uniform nanostructured materials cannot be prepared through manipulation typical of the design on macroscopic scale as easily and efficiently as they can be prepared via relying on self-organization and self-assembly phenomena. And this is exactly where colloid science can advisably step in.

Chapter 2

THE PRINCIPLES OF COLLOID SCIENCE

Colloid science deals with multi-phase systems in which one or more phases are dispersed in a continuous phase of different composition or state. Dispersed particles and dispersion medium can be in solid, liquid or gaseous forms, with the dispersed particles having typically between 1 nanometer and 1 millimeter in size. Contrary to ordinary solutions, colloid systems are thermodynamically unstable and only due to the existence of large interfacial energies that are stronger than thermal energy, kT, their order is preserved [1]. On the other hand, colloidal particles exist in a size range where the energy of random collisions with solvent molecules is adequate to maintain their distribution throughout the medium against the sedimentary action of gravity. Table 1 presents examples of different types of colloidal dispersions, divided according to the specific molecular aggregation states of the dispersed and continuous phase.

Although some of the practical principles of colloid chemistry were used in many ancient cultures for controlling inter-particle interaction energies and obtaining high-quality suspensions and final products in forms of inks and porcelains, Michael Faraday is said to be the first to define a procedure for stabilization of gold colloids in aqueous media through charging the particle surfaces and thus overcoming the tendencies of attractive van der Waals forces to clump the particles together [2]. The principle knowledge on the effects of various agents on colloid stability, mostly based on the simple framework of DLVO theory, is today routinely used in achieving and maintaining stability of colloids throughout numerous industrial sectors (foods, pharmaceutics, cosmetics, photography, electronics, agrochemicals, paints, building materials, etc.) as well as in desired disruption of stability of certain colloids by inducing flocculation and sedimentation (or creaming) of the constitutive particles, as in the cases of water purification.

Table 1. Different types of colloidal dispersions, divided according to the specific molecular aggregation states of dispersed phases and dispersion media.

Dispersed phase	Dispersion medium	Type	Example
Solid	Liquid	Sol	Paint, ink
Liquid	Gas	Aerosol	Fog, sprays
Solid	Gas	Aerosol	Smoke
Liquid	Liquid	Emulsion	Milk, blood
Liquid	Solid	Gel	Pearl, opal, cheese
Solid	Solid	Solid sol	Ruby glass, alloys
Gas	Liquid	Foam	Fire extinguisher
Gas	Solid	Solid foam	Pumice stone
Macromolecular particles	Liquid	Latex	Sap

DLVO theory, developed by Derjaguin, Landau, Verwey and Overbeek in the 1940s, rests on the assumption that the stability of a colloid dispersion is maintained through a balance between the repulsive electric double layer forces and the attractive van der Waals forces. In assuming so, it deploys the standard Gouy-Chapman-Stern concept of double-layer of charged species that surrounds each of the particles constituting certain colloid system. Since the stability of colloid systems inherently depends on the characteristics of the interface separating the dispersed and continuous phase [3], most of the effects used in controlling the colloid properties rest on the simple principles that involve modifications of the particle-medium interface by varying pH value, ionic strength, particle sizes and chemical agents that promote steric interactions. For instance, pH changes lead to modification of surface charge of the particles, increasing the surface charge density (or zeta potential) as introduced pH changes move away the system from its isoelectric (zero-charge) point and decreasing surface charge density in the opposite direction. Increases of ionic strength via the addition of salt can be used to screen Coulomb repulsion and compact the double layer that surrounds each of the charged particles. On the other side, as concentration of ionic species is reduced, the electrical double-layer thickness increases, and thus the range of the repulsion force increases. Reducing the size of colloid particles may increase the stability of the system by increasing the range of Brownian motion which overcomes the tendency of collided colloid particles to associate, grow in size and eventually segregate as a separate phase. Adding stabilizers such as high molecular weight polysaccharides, increasing the viscosity of the system, reducing particle movements and slowing down the slow but inevitable phase segregation processes, is regularly used for extending the shelf life of colloidal foods.

Yet amphiphilic additives, such as surfactants, lipids, copolymers and proteins, can self-assemble into a wide range of ordered structures and transform among various structural forms depending on the slight changes in solution or environment conditions. The versatility of potential colloid forms naturally reflects a wide range of potentially existing metastable equilibrium states. Microemulsions [4,5,6,7], clear and isotropic liquid mixtures of oil, water and surfactant that, unlike emulsions, do not require high shear rates for their formation may, for instance, typically exhibit a wide range of inherent multi-molecular configurations [8], including either regular or reverse micelles of various oval shapes (spherical, cylindrical, rod-shaped), vesicular structures, bilayer (lamellar) and cubic liquid crystals, columnar, mesh and bicontinuous mesophases, cubosomes (dispersed bicontinuous cubic liquid crystalline phase), sponge phases, hexagonal rod-like structures, spherulites (radially arranged rod-like micelles), multiphase-substructured configurations (such as water-in-oil-in-water or oil-in-water-in-oil droplets, for example), highly percolated pearl-like structures, supra-aggregates that comprise various substructured combinations of microemulsion aggregates, as well as numerous transient configurations. Each different point in a microemulsion phase diagram thus corresponds to specific local conditions for physico-chemical transformations that take place therein and result in unique material structures templated in each of these cases [9].

Besides displaying a rich phase behavior and a large structural diversity, the time scales for typical phase transitions in amphiphilic colloid systems can range over several orders of magnitude [10]. Whereas diffusion-controlled molecular rearrangements occur in sub-μs range, slow rearrangement of larger amphiphilic structures, such as in the cases of equilibration of vesicles or aggregation processes in surfactant – protein mixtures, controlled by weak attractive interactions, may take up to several weeks or months [11]. Although metastable amphiphilic systems are being given larger attention in the aspect of application of colloids, transient colloidal configurations can also be significant from the practical point of view, mostly as dynamic self-assembly 'templates' for the preparation of novel material structures. Just like fine variations in the effective colloid structures can give rise to significantly differing outcomes, relatively short-lived, non-equilibrium intermediate structures or microscopic phase separations may uniquely determine the final structure and properties of a material [12]. As a matter of fact, variously driven, dynamic and dissipative assembly phenomena wherein temporal stabilities of the system are maintained due to dissipating energy are abundant in Nature and have with many reasons been suggested as the next big frontier in the area of colloidal self-assembly [13].

Chapter 3

DICHOTOMY OF HARD-TECH VS. SOFT-TECH APPROACHES

The concept of perfect design seems to be inseparably linked to a perfect control over formation of inherent boundaries and differences within material structures at atomic and molecular, nanosized scale. Whereas the traditional methods of processing material structures, mostly based on employing intense pressures and high-temperature processing, prove to be inconvenient for controlling structural properties on small scale, the current generation of practical scientists and engineers is faced with challenges of developing the production processes that will at the same time be energy-efficient, low-temperature based and environmentally-friendly, and will be providing improved control over fine material structures, aimed at attaining superior performance characteristics. Miniaturization of electronic devices, evident since the creation of the first solid-state semiconductor device, has been performed by mechanically and chemically breaking down the materials to ever smaller sizes, which requires intense energy input as the scale of materials becomes ever smaller, but at the same time leads to uncontrollable fine structural features [14].

When it comes to evaluation of future prospectives for ultra-fine design of novel materials and technologies that may satisfy all of these requirements, two general lines of progress are usually depicted [15]. The one belongs to a hard-tech approach [16], which pertains to employing massive and complex instruments for manipulating atom by atom, molecule by molecule, or any other relatively simple building blocks, and organizing them into applicable structural outcomes. Such an approach to ultra-fine design is today frequently found under the name of 'molecular assembly', as Eric Drexler proposed within one of his visionary formulations of nanotechnological promises [17].

The essence of the second, soft-tech approach, that the practical scope of colloid science belongs to, is related to application of self-organizing phenomena of certain physico-chemical systems. In these processes, atoms, molecules and particles spontaneously arrange themselves into frequently sophisticated functional structures as driven by the energetic state of the system. The design of complex building blocks thus relies on initiating self-assembling, spontaneous and reversible processes by performing manipulations on macroscopic scale [18]. Such obtained products may have the potential to serve in further processing of more complex and eventually applicable functional structures.

Some of the strongest advantages of the soft-tech approach are in the use of relatively inexpensive, ordinary and commonly accessible processing equipment. Contrary to the hard-tech approach, it may in fact naturally induce prosperous decentralization of technological powers and sustainable global proliferation of practical knowledge. Instead of emphasizing oversimplified, 'Lego-like' depictions of creative manipulation of matter, accentuation of the importance of fundamental, basic science researches, being the root of every successful development and transfer of advanced technologies in the global society [19], prospers from the latter perspective. However, whereas the hard-tech approach to design of ultra-finely structured products may be seen as ignorant to natural limits and pathways, self-organizing, soft-tech approach typically lacks subtle abilities of tailoring the synthesized building blocks into functional appliances. Because soft-tech colloidal syntheses are implicitly based on wet processing, integration of such obtained self-assembled patterns onto the corresponding device substrates during its fabrication faces the problems of contamination, which has actually so far prevented soft-tech processing from becoming a viable option in the industrial fabrication of microtechnologies [20].

On the other side, just like a crane cannot be used for playing dominoes, it is natural to expect that relying on the same methods that are used for fulfillment of macroscopic design blueprints will not hold as a proper approach on a smaller scale. The so-called 'sticky fingers' problem [21] therefore presents one of the central obstacles that current hard-tech ideas are facing. It is a big question whether designing smaller and more precise manipulating tips, but essentially still relying on the same methodology that has been proven as efficient on macroscopic scale would present a successful approach. In that sense, Richard Smalley, Nobel Laureate in chemistry and one of the proponents of the soft-tech approach, has to say: "Much like you can't make a boy and a girl fall in love with each other simply by pushing them together, you cannot make precise chemistry occur as desired between two molecular objects with simple mechanical motion along a few degrees of freedom in the assembler-fixed frame of reference. Chemistry, like love, is more subtle than that. You need to guide the reactants down a particular reaction coordinate, and this coordinate treads through a many-dimensional hyperspace" [21]. As a matter of fact, it has been argued that since in order for a chemical reaction to occur not only directly interacting entities need to be bumped into each other, but - since electronic clouds that 'glue' chemically bonded atoms to one another are not local to each bond but sensitive to physical changes in the immediate surroundings - the overall spatial context of the reaction that tends to be initiated need be precisely set in order to obtain the desired reaction outcome [22]. If hoping to solve this problem by introducing more than one manipulating tip for each of the entities (or relatively small clusters thereof) of the near vicinity of the reaction, then it might be realized that there is not that much 'room at the bottom' as Richard Feynman proposed in his famous talk [16]. Every biological process organization in Nature - from enzymatic recognition processes to genetic evolution to human physiological functions [23,24], stochastic mental processes [25], species population changes [26] and even the logic of scientific discovery [27,28] - is founded upon immanent harmony between periodic order and chaotic randomness. Therefore, it is an important issue whether the design of novel materials and technologies in the coming era could transcend this compromise which today seems like an inescapable natural necessity.

Besides all mentioned, it is known that particle-like properties of macroscopic entities increasingly cede their place to quantum and surface effects as fine dimensions are reached. Contributions from interfacial regions play a dominant role in defining the effective properties of colloid systems, and wave-like properties at sufficiently small scale become as important as particle-like properties, suggesting certain invalidness of the 'block-building' approach below particular sizes. With decreasing the particle sizes, particle-particle interactions that are not desirable from the point of view of manipulation of individual particles also become dominant in comparison with the effective influence of external force on individual particles [29]. Spherical particles having 3 nm in diameter possess approximately half of its constitutive atoms positioned on the particle surfaces, often drastically increasing reactivity of the system and making it significantly different comparing to the physical behavior of its bulk counterpart. The simplest imagined manipulation of such small particles may be their dispersion and formation of isotropic sols. However, poor contemporary knowledge on the interface effects leads to the fact that quite frequently more than 90 % of finely dispersed systems consists of a given steric agent, such as a surfactant or a polymer compound, which significantly influences certain additive physical properties of the system. An illustrative example may be found in numerous cases of preparation of superparamagnetic fluids [30,31,32], where already decreased magnetization of the particles (due to largely amorphized and magnetically disordered arrangement of individual magnetic moments at the particle surfaces) is further effectively decreased due to a high proportion of stabilizing additive that keeps the highly reactive particles from aggregating and forming one giant multi-domain magnetic phase. Most of the achievements in the field of stabilization of magnetic fluids have come through a subtle combination of general principles in steric repulsion with numerous manually-correcting trial-and-error preparation loops.

Mesoscopic systems, comprising typical boundaries in the size range of tens of nanometers to hundreds of micrometers, are positioned right at the interface between classical mechanical and quantum mechanical behavior, making the statistical many-body theory practically inapplicable in this realm [33]. On the other hand, comparing to the scientific areas oriented towards investigating phenomena on either extremely large-scales (i.e. astronomy, ecology) or extremely small scales (i.e. quantum mechanics, high-energy particle physics), the subject of colloid science is dominated by phenomena and size ranges that could be conveniently treated both theoretically and experimentally. However, the complexity of the fundamental colloid science can be partly ascribed to the problems involved in describing assemblies of small particles adequately in thermodynamic terms, and partly to interrelation of many effects and parameters involved in determining the actual forms of energy surfaces and the mechanisms of its dynamic influence on the states of the whole system. Nevertheless, a remarkable advance in the techniques for manipulation and dynamic shape transformation of colloidal building blocks, such as via implementation of optical [33,34,35] and magnetic [36] tweezers, electrokinetic (electrophoresis, dielectrophoresis, travelling-wave dielectrophoresis) and radiation pressure forces [37], ultrasonic radiation [38] and acoustic traps, laser beams [35,39], hydrodynamic flows and atomic force microscopy [21,40], is noticed. On the other side, self-assembly phenomena, previously limited only to biotic examples, such as protein folding mechanisms and nucleation of inorganic crystals found in sea shells, are now being fruitfully explored in laboratories world-wide and routinely used for controllable production of many novel microphase textures [41,42]. The future prosperity in the design of novel and advanced functional supramolecular structures may correspondingly be present in the complementarity of the so-called 'top-down' and 'bottom-up' ultra-fine design methodologies. Nevertheless, optimistic beliefs in both endless possibilities of sole molecular machining approach [43], and in potential fruitfulness of self-organization of relatively large and complex structures into hierarchically structured outcomes, are widely present today [44].

Even though it may look that biological design relies on seemingly unpromising soft materials, such as proteins, lipids and polysaccharides, as well as on random design methods that are restricted to the accidents of evolution - a remarkable trial-and-error optimization tool of Nature - and that hard materials like diamond tips would prove as more efficient in fragmenting and repositioning individual atoms, molecules and their clusters into sophisticated nanoscale machines, it may soon become evident that billions of years of evolution have in fact already yielded the contemporary biological structures as perfect nanomachines of the current evolutionary era [45], capable of subduing and efficiently using the very nanoscale effects that have recently been indicated as the inherent obstacles in the high-tech, mechanical approach to fine structural design: low Reynolds numbers (implying increased influence of viscosity, which becomes the source of major forces opposing the motion at sufficiently small dimensions, instead of inertia), ubiquitous Brownian motion and strong surface forces [46]. Achieving motion through molecular shape changes, altogether with spontaneous assembling of complex mesoscopic structures out of recognition effects and subtle correlations or constrains at molecular scale present some of the obvious features of biological design, without comparable inherent flexibility in either traditional macroscopic engineering or the product visions of molecular machining design.

Contrary to the hard-tech approach to design of small technologies that tends to subdue and direct natural pathways according to the imagined outcomes, colloid science is implicitly related to constructing fine structures in accordance with natural self-organization phenomena. In that sense, whereas hard-tech line of progress presents continuation of traditional ways that technological products are being designed (that is, by machining chunks of raw materials and their assembling), soft-tech colloid approach presents new and elegant ways for syntheses of both simple, uniform material structures and applicable hierarchical superstructures. Through self-organizing approach to ultra-fine engineering design, the doors to a novel attitude towards natural pathways open. Contrary to shallow approach (instigated by the proponents of hard-tech, socially and ecologically decontextualized design) according to which an epithet of 'natural' is deserved for all endeavours which do not violate fundamental physical laws of nature, a different outlook towards future design is being created, according to which only through innovations that preserve natural diversity of web of relationships in the living domain, is that truly long-term creative endeavours may sustain. However, such a design pervaded with an attitude of stewardship and wondering respect comes at the price of giving away a part of the designer's dreams for the way of Nature. Magician's wand that insensitively tries to make everything possible according to its own ideas is thus transformed into a creative and feedback-awakened attitude of being hand-in-hand with mother Nature, sensitively trying to balance our designer dreams with the way of Nature, and thus aiming for the right balance of creative ideas and natural self-organization pathways within every aspect of organization of life, from molecular to ecological levels.

Chapter 4

RE-VISITING THE PROMISES OF DLVO THEORY

DLVO theory [47,48] presents the cornerstone of a theoretical framework with immediate practical significance in colloidal materials design. In fact, the development of this theory was in large extent supported by industrial investments due to the expectations that its concepts might increase predictabilities in the field of practical colloid design [49]. However, despite the fact that it was developed in the 1940s, it has proved difficult ever since to refine it to the point at which, beside ordinary qualitative application, meaningful quantitative comparison with experiments would be possible as well. It is based on a statistical calculus and as such provides implicit limitations to its use in the so-called perfect design. Many approximations are necessarily presupposed within DLVO theory. First of all, the terms accounting for higher-order correlations among the charged particles are neglected. Implementing the mean field approximation yields the famous Poisson-Boltzmann equation, used as the basis for understanding electrolyte and macroionic behavior for more than 80 years [50]. However, by considering only one ionic distribution, the mean field approximation neglects fluctuations and higher-order correlations between the charged particles. Even as such, due to non-linearities associated with the mobile ion Boltzmann factors, Poisson-Boltzmann equation has no analytical solution except for the simplest, highly symmetric geometries [51], such as parallel charged plates are for example. The core point of DLVO theory is therefore invoking the approximations from the Debye-Hückel theory of electrolyte structure to linearize the Poisson-Boltzmann equation and enable its analytical solvability. However, the linearized model may be valid only for describing regions where the local potential is small comparing to the thermal energy scale, which does not hold near the surfaces of highly charged spheres, where non-linear effects are thought to be confined. On the other hand, for DLVO pair potential to be used in describing the many-body structure of a colloid suspension, another linear superposition - according to which the distribution of simple ions around one sphere is not disturbed by the presence of the other - needs to be invoked. However, these approximations limit the representational scope of the model to weakly interacting suspensions, i.e. only to the cases that comprise weakly charged spheres separated by distances that are much larger in comparison with the screening length of the system. Linear models for colloidal interactions in general fail to explain the elastic properties of strongly interacting charge-stabilized colloidal crystals, where many-body effects become important comparing to weakly coupled, diluted systems [52].

Even with excluding the non-linear terms, DLVO theory proves to be unsolvable for many-body systems. The interaction of three or more bodies necessarily produces non-linear phenomena which entail significant sensitivity upon initial conditions of the evolving system. It has been demonstrated that three identical spheres sedimenting through a viscous fluid in two or three dimensions undergo chaotic behavior [53]. Also, the addition of molecular force term in an equation for describing flow of a liquid film along an inclined solid surface yields chaotic fluctuations in the evolution of hydrodynamic stability of the film even for extremely small Reynolds numbers [54]. As a consequence, slight changes in the initial experimental settings can be reasonably expected to produce variety of interesting results. Such an analogy between complex colloid systems and feedback-permeated bioentities may be supplemented by few more similarities. For instance, comparing to thermodynamically stable ordinary solutions of chemical species, colloidal dispersions are - due to 'mixing the unmixable' - non-stable, irreversible and - because of the excess energy associated with the formation of new surfaces - require additional energy for their formation. In that sense, colloids may be said to present one of the first thermodynamic steps in the evolution of life from thermodynamically stable (equilibrated) matter to homeostatic biological systems. Irreversibility of colloidal dispersions makes them highly dependent on the way of preparation, in contrast with ordinary solutions, which is one more sign of the potential existence of significant sensitivity towards initial conditions within. However, whereas living organisms owe their homeostatic stability to a selective openness to the energy-flux of the environment, the very stability of colloid systems can be comprehended as merely a state wherein a specified process that causes the colloid to become a macrophase, such as aggregation, does not proceed at a significant rate.

That DLVO mean-field theory presents a hypothesis model proven useful in general prediction of behavior of colloid systems and not a perfect framework as well, presents the case of CAT theory (*Coulombic Attraction Theory*) [55]. Within the framework of this theory, developed by Sogami in 1983, the mean field interaction between charged particles has a weak, long-range attraction via their counterions [50,56], contrary to classical DLVO predictions according to which the long-range interaction among the dispersed particles is purely repulsive. The latter proposition (although it fails to describe the long-range repulsion of colloidal pairs) has been invoked in many recent cases for explaining a number of anomalous colloidal phenomena [57,58]. However, the explanation links that will intertwine the phenomenon of repulsion of isolated pairs of like-charged particles with the observed phenomenon of multi-particle attractions in cases when the density of the same spherical entities is increased, are still missing. Even suspensions with monovalent simple ions or low ionic strength frequently display puzzling phenomena from the perspective of DLVO theory [50,59].

Few more approximative assumptions within the theory may be mentioned. For instance, electrical double layer and van der Waals forces are basically treated independently and assumed additive [60]. Then, the boundary conditions for the double layer are constant charge or constant potential, which brings about additional averaging of dielectric constant over inner and outer Helmholtz planes. However, such an approximation particularly does not hold at high surface charge or potential. It is not only that molecularly smooth, solid and inert (except that they provide a source of counterions that leave behind charged surfaces) particles are presupposed within the theory (disregarding asymmetries in shape and size), but contact angles are also ignored. However, it has been acknowledged that there is no single mechanism that can account for the diversity of behavior observed with differently prepared hydrophobic surfaces [61]. Then, discreteness of charge effects is taken into account by a smeared out surface charge approximation, which has shown to be a solid approximation only at either very low or very high electrolyte concentrations, whereas it poses difficulties at intermediate ranges [60]. Also, assumption of a primitive model of electrolytes according to which point ions and additive hard core radii are invoked, and thus basically treating the double layer problem as a bulk electrolyte one has been proven as invalid on many occasions and is often corrected by employing fitting parameters that involve size of ions as well as by taking into account short-range interactions between ions in the electrolyte, and ions and the surface.

Many additional effects, so-called 'non-DLVO' forces, including hydration effects, Helfrich undulation forces, Casimir-Polder force, van der Waals-Lifschitz forces, and various structural, oscillatory, hydrophobic, depletion, solvation, fluctuation, and protrusion forces were proposed in order to consistently model the situations that standard, long-range DLVO theory apparently fails to describe. Fluctuations in the simple-ion distributions and surface charges and asymmetries that are omitted from the mean-field concept of DLVO theory, responsible for London dispersion elements in the effective van der Waals attraction forces, were suggested as the key to solving these problems [62]. As a matter of fact, when the charge is screened and the interparticle distance large, in the standard DLVO picture only Onsager-type transitions (that is the ones that evolve in the same way as deterministic perturbations do) or 'weak', long-range terms due to dipolar interactions remain as the most recognized driving force for self-assembly. Such a subtle forces, different from the interactions for which the driving force is tight packing or 'strong', short-range forces such as hydration, are known to present the key influences for explaining overwhelmingly large number of phase combinations that certain colloid systems can adopt [63]. However, combining the effects derived from various non-additive weak interactions - such as universally attractive dispersion forces, electrostatic forces that arise due to specific counter-ion distribution, hydration forces related to adsorption/desorption of solvent from interface, depletion forces that are of topological origin, and specific recognition interactions (such as between sugar, polypeptide or fatty acid active groups and moieties) - with the effects of flexibility and fluctuation of real dispersed particles in colloid mixtures [49], presents an enormous task for any attempt to comprehensively explain and systematically model intricate self-assembly phenomena at the colloidal scale. In that sense, the whole concept of 'mean field' proves reasonable in explaining only the suspensions with high concentrations of simple ions, when relative fluctuations in local simple ion concentrations are relatively small. Because the suspending fluid appears in this model only via its dielectric constant, with neglecting the effects due to molecular structure of the solvent, the supporting medium is in the framework of DLVO theory generally treated as a continuum, which it rarely is in the real case. The assumption that an intervening liquid has a uniform density and orientation profile, supported by pairwise summation of London dispersion forces, stands clearly against the experimental evidences according to which, instead of having bulk properties all the way down to 'contact' (as is assumed by the theory), water undergoes drastic modifications in nanodomain structure, molecular mobility, hydrogen bonding, relaxation processes and the overall solvent characteristics as it approaches foreign surfaces [64]. Although two major types of water structure and reactivity may be defined: a less dense water region with an open hydrogen-bonded network against hydrophobic surfaces, and a more dense water region with a collapsed hydrogen-bonded network against hydrophilic surfaces, diverse self-association mechanisms can result depending on finely localized solvent properties against hydrophilic and hydrophobic surfaces [65]. Not including any overlap of surface-induced order through either the effects of hydration, uneven density or molecular orientation of the solvent (thus opposing the Poison assumption) is implicitly equal to assuming that interfacial free energies can be taken as infinite, as they are unperturbed by interaction, which is an approximation with a wide range of limitation consequences. Such a primitive description of the solvent presents a source of serious inconsistencies that restricts the theory to describing phenomena at scales only larger than a few nanometers.

It is obvious nowadays that in order to maintain physical significance, the current status of DLVO theory necessarily needs to balance between floating in the solvable areas of large approximations and facing an unpredictable continuum of parameter space due to extending *ad hoc* decorations in the starting settings of the theory. Explaining the behavior of molecular assemblies within ever more complex colloid structures produced in advanced chemical laboratories out of formulations that satisfy ideal or extremely simple physical situations at a mesoscopic level, would definitely remain a future challenge on the part of colloid theorists. And on the other side, whether not only descriptive, but fundamental guiding of synthesis conditions by means of the conceptual framework of DLVO theory would be made possible in the future as well, so far presents only a distant dream on the part of colloid experimentalists.

Chapter 5

TRIAL-AND-ERROR AND BUTTERFLY-EFFECT EXAMPLES

When it comes to colloid science, it is obvious that theoretical models are still lagging behind experimental achievements. In fact, colloid science has been thriving from ancient Egyptian inks and clays to sophisticated pharmaceutical products mostly due to innovative engineering approaches, with only the modest assistance of available characterization techniques and theoretical modeling [1]. However, it has been acknowledged that quantitative and qualitative inadequacies in the linear theory of colloidal interactions may pose impediments to progress in wide range of fields, from protein crystallization to stabilization of industrial suspensions [52]. Numerous variables invoked for explaining physical properties of colloid systems at molecular level have remained unrelated to experimentally varied parameters of synthesis procedures. Table 2 presents parallel columns of macroscopic parameters modified in the regular design procedures and microscopic-sized variables and quantities that govern the self-organization processes in the observed and/or applied micellar systems.

Today's colloid experimentalists in most cases rely on theoretical knowledge in quite sketchy manner. That is, instead of attempting to correlate the parameters of experimental settings and outcomes with variables that are used to explain subtle, molecular effects within the models of a given theory, only basic phenomena are being invoked. These are for example:

a) pH- or particle size-dependent changes in potential energy barrier that keeps stable dispersions from flocculating;

b) contraction of charged layers surrounding individual suspended particles, leading to additional suppression of electrostatic repulsion at large distances by means of increasing the ionic strength of the dispersion medium;
c) addition of the proper surfactants (that are mostly chosen on the basis of their ready availability as opposed to a time-consuming structure-performance analysis [66,67]) or block co-polymers that would selectively induce either steric repulsion or flocculation, according to the desired outcome, i.e. dispersion or phase segregation.

Table 2. Macroscopic and nanoscopic variables in the microemulsion-assisted and particularly reverse-micellar synthesis of nanoparticles.

Macroscopic Parameters	Nanosized Parameters
Identity of included chemical species Microemulsion composition Water-to-surfactant molar ratio pH Ionic strength Dissolved species concentrations Method and rate of introduction of species Temperature and pressure Aging times Method and rate of stirring Homogenous or heterogenous nucleation	Static, size and shape distribution of micelles Aggregation number Dynamic interaction, rates and types of merging and dissociation of micelles Distribution of charged entities around dispersed particles Surfactant film curvature and head-group spacing Effective Coulomb repulsion potential Van der Waals, hydrogen and hydrophobic interactions Screening length

Variations of any one of the given experimental parameters necessarily produce influence on each one of the molecular variables of the investigated process. Likewise, all of the internal variables used for the fundamental description of a system are mutually intertwined, resulting in peculiarity of each specific system under the given set of conditions (composition, initial thermodynamic state, an interface with the environment). Whereas kinetic conditions in ordinary solutions may reasonably be approximated as continuous, dynamics of solvation effects and reaction kinetics can - depending on the structure of the microheterogeneous colloid system - largely vary in different local microenvironments, effectively producing significantly complex outcomes.

Whereas the dynamics of solvation effects can drastically change with an interfacial distance, in the cases of chemical reactions performed within micellar aggregates, the components of the reaction rate constant include the effects of Brownian diffusion of reverse micelles, droplet collision, water channel opening, complete or partial merging of micelles, diffusion of reactants and the chemical reaction, as well as fragmentation of transient dimers or multimers (wherein the slowest step determines the temporal aspect of the overall process of synthesis) [68], ranging from the order of magnitude of nanoseconds for diffusion-controlled intermicellar reaction to an order of miliseconds for intermicellar exchange of reactants [69]. Further more, despite the fact that dynamic response in colloid systems is typically much slower comparing to their bulk counterparts [70], extremely fast responses may be favorable under certain conditions, as can be illustrated by numerous examples of catalytic effects produced by the influence of micellar encapsulation [4,71,72,73,74,75] and exchange [69,76,77] of reactants. On the other hand, the rates of biomolecular reactions performed within amphiphilic media that comprise micelles, vesicles or other types of colloid aggregates are shown to vary from promoting highly catalytic to thoroughly inhibiting conditions depending on slight variations of concentrations and types of applied amphiphiles and salts [71]. Such variations may significantly influence counterion and co-ion distributions and therefore the subtle selectivity phenomena at the interfacial regions that govern the catalytic behavior of the given colloid systems.

Beside obvious intricacies, there were numerous attempts to correlate certain colloid media properties in an oversimplified and overgeneralized manner. In the field of reverse-micellar preparation of nanomaterials [5], for example, it was proposed that the size of particles obtained by precipitation in reverse-micellar microemulsions based on cetyltrimethylammonium bromide (CTAB) as a surfactant ought to be equal to water-to-surfactant molar ratio in nanometers [78]. Similar to this, Pileni proposed adding only a factor of 1.5 in the mentioned relationship in order to apply this correlation to microemulsion systems based on sodium bis(2-ethylhexyl) sulfosuccinate (AOT) as a surfactant [79,80]. Although this relationship was verified only for certain compositions of specific AOT-based microemulsions and particles prepared within [80], it has been frequently mistaken as corresponding to all types of microemulsions and particles [4]. As a response to such an oversimplification, numerous cases of experimental deviations from the proposed correlations were reported [81,82,83,84]. It is not only that water-to-surfactant molar ratio in reverse-micellar ranges of the given microemulsion phase diagrams does not correspond to micellar sizes in direct proportion in all cases [81], but the very same small-angle X-ray scattering

(SAXS) characterization technique that was relied upon in defining the mentioned relationship between water-to-surfactant molar ratio and the size of produced particles [80,85,86,87], has shown that micellar radii in the same AOT/isooctane/water system change in response to an addition of small amounts of compounds solubilized in the microemulsion [79]. Experimental results indicate that size of reverse micelles depends not only on water-to-surfactant molar ratio, but also on identity of all the included microemulsion components, their respective concentrations, pH, temperature and ionic interactions caused by introduced electrolytes or immanently dissociated molecular species [5]. Also, the particle formation processes necessarily affect structure of the parent emulsion, resulting in a feedback interaction that ends as either a form of phase segregation or a metastable state in cases when isotropic colloidal dispersion structure is preserved. It has been known that phase diagrams of microemulsions derived with and without the presence of the prepared material or any other additional component may be drastically different [14]. Therefore, in light of such mutual transformations, the whole concept of 'templating' as the translation of the shapes and sizes of the self-assembled organic species onto the structure of the nucleated and grown crystallites, looks as if it needs to be reevaluated, particularly in the area of reverse micellar preparation of materials where the phrases like 'nano-cages', 'nano-templates' or 'nano-reactors' seem to dominate the explanations of particle formation mechanisms. For the most surfactant-mediated syntheses, connection between morphology of the surfactant aggregates and the resulting particle structure is more complex (than simply relating the average size and shape of the micelles to size and shapes of the precipitated particles) and affected by hardly reducible conditions prevailing in the local microenvironment that surrounds the growing particle [10]. Composition, pH, concentration of the reactants, ionic strength and heat content are some of the experimentally modified variables that co-influence this local environment. As chemical reactions and physical transformations caused by aging take place within a colloid and its corresponding microenvironments, many of these factors are liable to change. The decoupling of effects that belong to each specific macroscopic modification of the system presents one of the biggest challenges in the practical field of colloid science.

Another oversimplified idea in the area of reverse-micellar assisted preparation of nanoparticles was that size of the produced particles was supposed to be equal to the size of the micelles that cap and limit the growth of individual crystallization nuclei. Despite such a picturesque representation of the processes of particle formation inside the so-called nano-reactors (i.e. 'water pools') of reverse micelles, numerous cases wherein the variations in the produced particle sizes could not be correlated with sizes of the reverse micelles, were reported [81,82,83,84,85,88,89]. However, as was previously mentioned, the size of the multimolecular colloid units or any other relevant property of a colloid system cannot be dependent upon any single internal variable, but only on the complex interactions that are conditional for its existence. Many cases support the idea that the reasons for the frequent mismatch of the properties and quantities derived by different experimental methods, and that particularly when investigation of micellar properties is concerned, do not result from errors inherent in the experiments, but is evidential of a fundamental shortcoming in the single parameter models [90]. Despite this, whereas on one hand such one-parameter models are still routinely used to predict micellar and vesicular sizes [91], poor current knowledge of the underlying mechanisms that govern micellar-induced synthesis processes and impossibilities to *a priori* predict the distributions of sizes, shapes and the effective physical properties of the processed systems is acknowledged on the other hand [92]. Such a situation is highly reminiscent of numerous oversimplified attempts to infer hydrophobic interactions from molecular-scale surface areas alone, even though bulk driving forces and interfacial effects compete in determining hydrophobic effects in any particular case [93]. As a more reasonable explanation, the dynamic interaction among colloid aggregates has since lately been generally considered as the most important factor that influences the morphology and the properties of the final reaction products [69]. However, since dynamic interaction of colloid multi-molecular aggregates, such as micelles, cannot be yet directly observed in real-time conditions, indirect techniques are usually applied in order to evaluate both static and dynamic properties of the corresponding media. In the approximations (introduced in order to overcome the limitations of characterization techniques in terms of sampling, experiment time scales, etc.) and different implicit presuppositions of various such techniques are present the reasons for a frequent mismatch [94,95] between the concluded properties attributed to the same systems by using different experimental methods. Trial-and-error, i.e. ‘tapping in the dark’, or state-of-the-art as pessimists and optimists sometimes refer to it, respectively, thus presents the core of approaches to obtain novel outcomes of practical significance in the field of today's colloid science.

As far as the current state of the art is concerned, it is exceedingly difficult to predict the outcomes of experimental settings aimed to produce novel fine structures and morphologies by the means of colloid science methodology, and the most attractive results in today’s practical colloid science come from trial-and-error approaches. There are many evidences that slight changes in the limiting conditions of particle synthesis experiments can produce significant differences as the end results [5]. The following examples may illustrate such a proposition and enrich one's belief in crucial sensitivity and subtleness of the material design procedures that involve wet environments and colloidal phenomena.

Replacement of manganese ions with nickel ions in an experiment of reverse-micellar precipitation synthesis of the mixed zinc-ferrite resulted in the production of spherical particles in the former case [96] and acicular ones in the latter [97,98]. When bromide ions of cetyltrimethylammoniumbromide (CTAB) surfactant were in a particular sol-derived synthesis of barium-fluoride nanoparticles replaced by chloride ions (CTAC), identity of the final product was no longer the same, whereas a replacement of 2-octanol with 1-octanol significantly modified crystallinity of the obtained powder [89]. Whereas porphyrins substituted with two cyanophenyl moieties in a *cis* configuration gave rise to compact clusters that comprised four molecules each, the respective substitution in a *trans* configuration led to self-assembly of linear molecular chains [99].

When a microemulsion-assisted synthesis of copper nanocrystals was performed in the presence of sodium fluoride, sodium chloride, sodium bromide or sodium nitrate, small cubes, long rods, larger cubes and variety of shapes resulted, respectively [100]. Variations of salt identities and concentrations in another case of colloidal preparation of copper nanocrystals also resulted in drastic morphological changes [101]. The drastic effects of electrolytes on colloid properties were acknowledged in the past [102], and whereas the effects of Hofmeister series of anions on protein solvability and a wide range of other phenomena still remain unexplained by the present theories of physical chemistry [103], classical DLVO theory treats ions only through their valence [104], disregarding numerous other physical qualities, including short-range interactions.

Various choice of precipitation agents within *in situ* preparation of colloid particles from the solutions can often result in distinctive morphologies obtained [105,106]. As a matter of fact, the solvation forces that are sensitive to slightest changes in particle size, shape, surface roughness, particle-solvent interactions and solvent structure, can in large extent influence colloid particle interactions, including their proposed role in self-alignment of nanoparticles in solution [107].

Slight changes in micellar dispersity towards wider polydisperse distributions may trigger the processes of Ostwald ripening of the colloid particles that result in the phase segregation [90]. Sequences of minor changes in initial conditions for a seeded agglomeration of sodium aluminate particles from liquors are shown to be enlarged during the process, resulting in aperiodic oscillations in particle size distributions [108]. In a colloidal process of preparation of silver particles, slow rate of the addition of reducing agent resulted in formation of spherical morphologies, whereby an increased rate of addition led to obtaining platelet-shaped silver particles [109].

Prediction of self-assembled patterns of colloid particles is often non-trivial due to complex competition between various attractive and repulsive forces of different ranges that governs their multi-scale clustering arrangement. A well-known example comes from the ordering of spinning particles by the influence of magnetic rotating field (i.e. the spinning rate of a magnet) wherein interference of vortices produces chaotic conditions, resulting in inability to predict numerous attractive patterns as outcomes of the self-assembly processes [110]. The order in vibrated granular layers is also shown to appear in the most unexpected circumstances, depending only on the volume fraction of rods in a vibrating container [111], and numerous other examples may be found as related to competition between chaotic mixing and ordering segregation tendencies in the field of granular coarsening dynamics [112]. An example wherein increasing the attraction between particles in a glassy colloid system unexpectedly breaks up its glassy structure shows how macroscopic physical properties of colloidal suspensions can be dramatically influenced by tuning the interactions between their constitutive particles [113,114]. It is important to note again that this interaction cannot be directly influenced, in a sense that experimentally controllable parameters (such as external fields strength, its speed of rotation or speed of the cylinder in the examples given in this paragraph) may be controlled with high fidelity, but not the qualitative nature of the interactions between the particles that would allow for perfectly precise introduction of coding instructions into the elements of the system and consummate exploitation of self-organization phenomena for practical uses [115].

As far as the behavior of colloids in environmental context is concerned, it can be noticed that although during the water purification treatments, trace elements are usually bound to colloids, values of 0 – 100 % of binding have been observed depending on the element that is measured, the analytical technique that is employed, the type of water, and other factors [116]. Regarding the biological context of application of material products, the comparison between harmful serpentine, layered mineral form of asbestos and structurally different and highly toxic, fibered modification of the same chemical compound, presents a classic example of drastically different outcomes upon seemingly slight variations of a material properties prior to its environmental introduction and wide application. Whereas it has been known on one hand that the compounds that possess therapeutic effects may turn toxic when delivered in the same chemical, but different crystalline form, relatively minor changes in the preparation conditions on the other hand may result in drug particles of the same composition, but of various shapes, as was evidenced in the case of colloidal re-crystallization of loratadine and danazol [117], naproxen [118], calcium and barium naproxenates [119], hydrocortisone 'budesonide' [120] and cyclosporine [121] particles, eventually produced in wide ranges of crystalline forms, including fibers, rods, rectangular and plate-shaped particles and uniform spheres. It has also been known that apparently minor changes in molecular configuration of certain catalytic compounds, their microstructural arrangement or interfacial structure may possibly lead to significantly modified resulting catalytic properties.

With decreasing the particle size of a given material towards nano-dimensions, numerous novel physical properties may appear, significantly differing from the results of approximations or simple extrapolations of physical features of the bulk materials with identical composition down to small sizes. Novel mechanisms of explanation of behavior of small particles comparing to their micro-sized counterparts are necessary in such cases. Few examples may be offered to illustrate such drastic modifications of physical properties. Gold as an inert and unreactive material in bulk atomic symmetries becomes highly efficient catalyst as the transition to nanoscale order is introduced, whereby silver nanoparticles exhibit bioactive properties absent from larger particles [122]. Soft and malleable carbon in the form of graphite can become stronger than steel, whereby aluminum may spontaneously combust and even be used in rocket fuel with the transition to nanoscale [123]. Metals that are malleable when they are microstructurally arranged may prove to have unacceptable levels of creep when their grains are reduced to nanolevel, whereas formability of ceramics (that are typically hard, stiff and brittle) is known to improve with the reduction of grain sizes towards nanoscale, so that certain ceramic materials, such as Y_2O_3-stabilized ZrO_2, liquid-phase sintered Si_3N_4 or even SiC, may above approximately half the absolute melting point exhibit superplastic behavior [124]. Since mechanical failure of a material most frequently takes place through crack migration processes along grain interfaces, the fact that materials with nanosized grains (down to ~ 10 nm) are stronger comparing to their bulk counterparts implies significant modifications of strength and toughness mechanisms with the transition from bulk to nanoscale simulations [125]. The classical Hall-Petch model generally fails to fit experimentally observed hardness vs. grain size dependencies at sufficiently small grain sizes that pertain to nano range (typically < 10 nm), whereas additional phases in complex composite structures frequently change the mechanical properties of the former nanocrystalline matrices in an unexpected way, with often exhibiting sudden slopes in some of the mechanical qualities at certain subtle, only few nanometers wide particle size ranges [126].

It has frequently been observed that insoluble substances become soluble, or that insulating compounds become conductive when their constitutive particles are reduced to nanosizes. Ultra-small capacitors at fine grain boundaries might be charged with even single electrons, strongly influencing the subsequent transport of charges through the material [125]. Because liquid phase at nano-dimensions has lower surface energy than a solid with facets, edges and corners, solid particles with few nanometers in size can frequently display twice lower melting temperature comparing to their corresponding bulk solids [127]. Although experimental evidences have been leading to enormous amount of data on changes with micro-to-nano transitions in many particular systems, there is still no theoretical scheme according to which it would be possible to predict how other materials, let alone materials in general, would behave when dispersed in nano ranges. Due to complex interplay between surface and quantum effects, each nanostructure can be led to display an array of unique potential properties with only slight modifications in their structural features, so that special discussions are obviously needed for each particular nanosystem, and that at all levels of their investigation, from chemical pathways of synthesis to the ranges of physical features to their environmental influence [15].

Since colloidal suspensions frequently comprise nano-sized particles, the uncertainties related to micro-to-nano extrapolations might be implied in cases that comprise colloidal systems as well. Three main characteristics that differ nano-sized aggregates comparing to larger, classical ones, including: i) confinement effects (i.e. quantization of energy states); ii) increased grain boundary interface; and iii) reaching the critical dimensions for certain physical effects, may be intermingled in variety of complex ways depending on sizes and shapes of the particles and their interaction, so that it would be exceedingly hard to predict physical properties of a system based on postulating its mesoscopic structure. Polygonal particles, fibers, layers, intergranular films, disc-shaped, acicular, lamellar, cylindrical particles or specific bicontinuous structures may, for instance, reach critical dimensions in only one direction, resulting in a peculiar behavior, often hard to predict by the conventional means [128]. Further on, key patterns in explaining the resulting physical properties in not only hierarchical, multi-scale structure of some of the most complex colloid systems, but in relatively simple composite materials as well may span 15 orders of magnitude, ranging from electron orbitals overlap to atomic, dislocation and grain microstructure to component arrangement to their complex overall organization within a single functional whole, making the perfect prediction of the full range of their behavior a distant [126] if not completely unattainable aim [129].

Therefore, it is not only complex chemical pathways, like photosynthesis or corrosion [130], that still present enigmas to scientific minds, but even static material structures may pose unsurpassable obstructions on the way towards their understanding. Rare-earth doped manganites wherein the complex competition between ferromagnetic metallic and antiferromagnetic insulating phases outlines the resulting electrical and magnetic properties, may present a neat example at this point [131,132,133,134]. Even though double exchange interaction between Mn^{3+} and Mn^{4+} ions was believed to solely define the physical properties of perovskite manganites, it has become obvious that a model based on that cannot be sufficient to explain the whole range of behavior of these materials. It is nowadays clear that in addition to including superexchange interaction in $Mn^{3+} – O – Mn^{4+}$ ion system and the corresponding Jahn-Teller effect, spin lattice or charge lattice interactions, i.e. polaron effects, need to be taken into account in explaining the physical characteristics of these compounds [135]. As far as the practical, synthesis-oriented perspective is concerned, this means that small variations in stoichiometry or particularly $Mn^{3+} : Mn^{4+}$ ratio, as well as in the microstructural features (defined by the type and conditions of a given procedure of preparation), and that particularly at nano-scale, including the thin films thereof (that may exhibit colossal magnetoresistance effect), may lead to achieving unexpected attractive resulting properties of the material. According to the paradigmatic sequence of material science domains of research interest wherein processing defines structure that defines properties that defines performance [136] (although at this place, a closed, iterative loop would be more appropriate, because material performance in form of novel technologies obviously modifies the design and inventory imagination that guides the processing methods [15]), such a complex possibility space of interactions between various atomic interactions and structurally-defined symmetries as in the case of manganites is naturally reflected on a wide array of both potential and actual application of this type of material. Only lanthanum-strontium manganites may exhibit colossal, giant or low-field magnetoresistance (due to unusual magnetotransport properties), high electrical conductivity, electrical field polarizability, half-metallicity of electron bands, anisotropic superexchange interaction due to Jahn-Teller effect, exceptional magnetoelastic properties, large sensitivity of interaction between conductive and insulating phases with temperature (with the typical transition at Curie point) and have therefore been used in magnetic refrigerators, as cathode material for solid oxide fuel cells, as magnetic sensors, as read heads in magnetic storage media, as parts of magnetic tunnel junctions, as catalysts in waste gas purification and catalytic combustion processes, and have recently been proposed by the author's research group [137,138] as a convenient material for hyperthermic and drug delivery biomedical treatments due to its stoichiometrically dependent Curie point.

The following examples may illustrate the idea that often routinely neglected influences in the preparation procedures may leave significant traces on the properties of the final products. It has been found that even the method of stirring in some of the microemulsion-assisted procedure of preparation of nanoparticulate colloids can have decisive influence on some of the final particle properties. Thus, using a magnetically coupled stir bar during an aging of a dispersion of particles influenced crystal quality and in some cases resulted in a different crystal structure as compared with non-magnetically agitated solutions [139]. In case of a synthesis of organic nanoparticles in reverse micelles, the use of magnetic stirrer led to the formation of nanoparticles larger in size comparing to the particles obtained with using ultrasound bath as a mixer, even though no changes in particle size were detected on varying solvent type, microemulsion composition, reactant concentrations and even geometry and volume of the vessel [140].

Changes in the sequence of introduction of individual components within a precursor colloid system could result in different properties of the final reaction products [141]. Changes in size of a volume where the particle preparation processes take place - as occurs when the transition from small-scale research units to larger industrial vessels is attempted - can lead to extensive variations in some of the properties of the synthesized material [142]. For instance, absorptivity of cadmium-sulfide particles has been dramatically changed when the amounts of the microemulsions used in the synthesis procedure were tripled [143]. Also, when two same chemical procedures for a colloid synthesis of nanoparticles were performed in closed and open, otherwise identical vessels, perfectly uniform spherical particles were yielded in the former case, whereby elongated particles of similarly narrow size distribution were produced in the latter [142].

An array of the presented cases wherein subtle modifications in the initial experimental conditions induce significantly peculiar outcomes can clearly lead one to a conclusion that complex synthesis pathways based on colloid, thermodynamically unstable systems, as in large extent opposite to relatively simple traditional, solid-state routes, offer inexhaustible potentials for single aimed chemical compositions. Attractive morphological, microstructural and hierarchical structural arrangements can sometimes be obtained by unexpectable and unpredictable settings of the starting and limiting conditions of experiments of synthesis.

Unlike some of the surfactant-templating synthesis that can be considered as structurally transcriptive (a copying or casting as in the cases of some porous inorganics [144]), ‘templating’ of crystallization processes within fine and sensitive, advanced colloid systems such as microemulsions can be regarded as first synergistic and only then as reconstitutive [145]. For example, although most of the particles produced in reverse micellar, AOT-based microemulsion systems were spherical in nature [146], crystallization of barium sulfate resulted in extended crystalline nanofibers aligned to form superstructures, whereby a precipitation of barium chromate in the same microemulsion system resulted in primary cuboids aligned to linear 'caterpillars' or rectangular mosaics [145]. Also, in the case of synthesis of calcium phosphates, variations in the relative concentrations of the microemulsion components resulted in numerous morphologies, ranging from co-aligned filaments to amorphous nanoparticles, hollow spheres, spherical octacalcium phosphate aggregates of plates, and elongated plates of calcium hydrogen phosphate dehydrate [147]. On the other hand, in the first historical report on the synthesis of materials in reverse micellar media [148], it was observed that size of the prepared platinum, rhodium, palladium and iridum particles was always in the range of 2 – 5 nm, independently on surfactant, water and reactant concentrations applied in the experiments [149].

Far from being only inert constraints to the growth of crystallites, microemulsions were shown to be physico-chemically active in defining the reaction pathways that take place in their presence, thus influencing the very chemical identities of the final products [130]. One of the causes for such observations may be present in significantly changed structure and physicochemical properties of interfacial water comparing to its bulk counterparts. The behavior of liquid molecules confined in nanosized spaces or at solid-liquid interfaces in general, due to surface-induced structuring, significantly differs from their behavior within a bulk system [150]. It has previously been shown that water molecules capped inside reverse micellar domains can be divided into three groups based on their mobility. FTIR studies suggest that the water interior of a reverse micelle has a multilayered structure, consisting of the so-called interfacial, intermediate and core water. The interfacial layer is composed of water molecules that are bounded directly to polar head-groups of a surfactant; the intermediate layer consists of the next few nearest-neighbor water molecules that can exchange their state with interfacial water; and the core layer is found at the interior of the water pool and has the properties of bulk water [151]. Depending on the size of reverse micelles, available water may have significantly different solvent properties, ranging from highly structurized interiors [152,153] to free water cores that approximate bulk water solvent characteristics. Different water structures may also dissolve different amounts of gases, which can drastically influence the reaction pathways, particularly in the cases where oxidation or reduction reactions by means of dissolved gases comprise crucial steps in the preparation procedures, as the numerous cases of ferric oxides may illustrate [130]. The accumulated gases are significantly present at hydrophobic interfaces comparing to the typical range of dissolved gas concentration in water at normal pressure and temperature ($\sim 5 \cdot 10^{-3}$ M) [60]. Fine variations in the experimental outcomes depending on the gas effects have been noticed[130], and there were cases where certain effects, which depended on many parameters, disappeared on removal of the dissolved gas [60]. Interfacial self-association mechanisms can also be quite different depending on the surface wettability. As a biological example, the rate of blood coagulation tends to increase with an increase in water-wettability of the tube surface [65]. Surface inhomogeneities in either chemical composition or morphology are known to play a crucial role in determining the kinetics and thermodynamics of wetting or dewetting processes, so that it has been noticed that a meaningful shift from typical theoretical treatments of liquid-solid interfaces as homogeneous and perfectly flat to more complex representations of dynamic and multiphase interfacial phenomena would require interdisciplinary research efforts that would involve the fields of physical chemistry (that invokes inherent electrostatic and van der Waals interactions), surface chemistry (oriented towards the phenomena arising out of the effects of interfacial free energies and surface structures), statistical dynamics (useful in explaining the effect of pinning of contact lines and transitions among wetting morphologies) and fluid dynamics (related to phenomena of liquid spreading and capillary flow) [154]. Also, self-assembly processes that occur during the drying steps of synthesis procedures involve complex competition between the kinetics of evaporation and the time scales with which solvated nano-particles diffuse on a substrate, and due to the specific role of hydrophobic interactions and a variety of ways to nucleate evaporation may lead to unexplored territories in the field of novel design [93]. As has been previously mentioned, treating water as a continuum medium and disregarding its fine interactions with gases, salts and electromagnetic fields may in future indeed cause ever increasing difficulties in attempts to explain fine variations from the ranges of expected results.

To generally conclude the ideas and observations presented in this section, we may be reminded that premature generalizations and oversimplified depictions of phenomena under investigation have indeed presented initial steps in the development of all areas of science throughout its history. However, only by acknowledging the relationship between natural phenomena and their scientific representations as not true reflections of objectively realistic systems, but as pragmatic metaphors (wherein map is obviously not a territory [25]) intended as ways of mutual coordination of human beings in the domain of our experiences, instead of constraining further evolution of understanding and enriching of the actual modeling schemes by regarding the proposed hypotheses and paradigmatic descriptions as universal and 'true' ones, an enormous potential space for their development would be opened. Both an attitude of seeking deviations from models without approaching the significance of their generality, and satisfaction with the existing models without seeking horizons where both acknowledgement of imperfections, uncertainties, limitations and impossibilities of scientific understanding and 'lands' of new discoveries are hidden, present incomplete approaches. The balance between the two – induction-derived generalizations and acknowledgement of their drawbacks and flaws - may be said to present to most fruitful basis for a scientific inquiry.

Chapter 6

THE SIGNIFICANCE OF PRACTICAL COLLOID SCIENCE IN THE LIVING DOMAIN

It has lately become clear that environmental context of application of novel materials ought to be necessarily considered in the very phase of their design in order to ensure durability and sustainability of both the materials and research efforts invested in their invention and production. The twelve principles of 'green chemistry' [155,156,157,158] and the ideals of employing 'natural surfactants' [159,160] present neat guidelines for increasing sustainability of both practical research results and the environment itself. Considering the fact that there is more than 1 million new chemical compounds reported each year and close to 1000 new ones being added to commercial domain every year (besides already over 70,000 existing) without any required training in toxicology or environmental sciences for doctoral chemists, it is clear that increasing ecosystem awareness regarding chemical life cycles would present a key component of 'greening' education in chemistry of the coming era [155]. On the other hand, learning from the synthesis pathways in biological realms provides numerous creative incentives for the field of modern colloidal design. For many millennia, natural relationships have provided thinkers, philosophers and scientists with metaphorical relations, copied as solutions for numerous scientific and ethical problems and dilemmas. Likewise, the imitation of natural syntheses, known as 'biomimicry' [161] approach, today presents a truly valuable basis for the design of advanced materials and structures. Since the cellular processes of synthesis occur in aqueous media with both solution and colloid properties (so that diffusion and interface effects seem to be intertwined), it is clear that with instigating biomimicry approaches, a transition from traditional, solid-state routes to an implementation of wet, colloid and soft-tech syntheses would implicitly take place. In fact, it has been suggested that since aqueous intracellular environment is rich with biopolymeric surfaces, understanding of sol-gel behavior would present a logical approach to an understanding of cytoplasmatic transportation mechanisms [162]. In that sense, Table 3 presents side-by-side typical investigated stimuli and the observed responses for some of the structure-sensitive polymer hydrogels. It is important to note that such phase transitions, corresponding to significant changes in structure and the effective physical properties that can bring about large-scale motions, can be triggered by subtle shifts in environment conditions [162]. Such a switch-like action with huge amplification is highly reminiscent of an organization of living systems, as described by the non-linear framework of complexity science [163,164].

Table 3. Typical stimuli and responses of polymer hydrogels [162].

Stimuli	Response
pH Temperature Chemical or biochemical agents Solvents Salts Electrical field Electromagnetic radiation Mechanical stress	Chemical (stimulates or inhibits reactions or recognition processes) Phase separation Shape change (shrinks or swells) Surface (changes in wetting properties) Permeability (sharp increase or decrease) Mechanical (hardens or softens) Optical (becomes transparent, turbid or changes color) Electrical (generates signal or electrochemical reaction occurs)

Similar to this, as the growth in complexity of individual organisms does not imply the growth of its cellular units, but the growth in complexity of their internal structures and interaction networks, sustainable production of the future could hardly be seen as one implementing gigantic production facilities at the cost of disappearance of smaller ones. A support for the development of complex patterns among smaller, ecologically more viable and more processing-efficient, 'meso-technological' [165] instruments thus may present an advantageous approach, inherent in the very foundation of the soft-tech concept of progress. Decentralized computational parallelism proves to be superior in solving complex mathematical equations to highly-centralized serial computers [166], a recent network graph model presented superior adaptive properties of chemical reaction networks (such as metabolic pathways, neural networks or planetary atmosphere) that exert short cycles and small world characteristics [167], and likewise, linking various production plants into 'symbiotic', zero-waste networks wherein (in accordance with Nature) the waste of one facility would present a resource to the other ones, presents the strongest nature-copying ideal on the horizon of global technological development. However, as the beginning of this section outlined, in order to direct the small design according to this aim, the largest possible context, including ecological, economical and social perspectives, needs to be steadily considered, which invokes a necessary need to awaken an inter-disciplinary eye, open to absorbing all sorts of knowledge and applying them on the actual, professional fields of interest. In that sense, the foundation of knowledge should really be based on the idea that "the challenge of the future may not be so much one of developing new theories and new hardware as one of developing better people" [168], as Guyford Stever, a former director of the National Science Foundation, once said.

Comparing to the synthesis processes in biological world that rely on using relatively simple building blocks, complex environments and complex processes, parallel processing (hundreds of reactions at a time), relatively slow reaching of final states, dealing with sub-picomolar quantities, and imperfect reproducibility (stochastically overcome by high selectivity for products that meet the required specifications), traditional synthetic methods in chemistry are typical of using relatively complex building blocks, simple media and simple processes, linear reactivity (one reaction at a time), quick reaching of final states (due to far-from-equilibrium conditions), dealing with molar quantities and tendencies towards duplicating reproducibility [169]. On the other hand, colloid science approach to preparation of novel, fine-structured materials often starts from simple building blocks that by aggregation form entities of more complex crystalline structures. Significant influences of aging times, viscosity bifurcation effect and a slow increase in critical stress values with the flow history and the resting time of a paste sample, typical of most thixotropic colloidal suspensions [170], indicate certain similarities with relatively slow (comparing to ionic reactions in solution) far-from-equilibrium reactions (due to numerous metastable states) that occur in the living domain [171]. As a matter of fact, depending on the non-equilibrium conditions of a colloid system, numerous attractive chemical environments that may specifically influence the growth of inherent mesoscopic structures can be produced, ranging from extremely fast responses that are deficient in intermediate metastable states (in that sense, we may recall that colloids present intermediate states of every phase transition between various aggregation states of matter, being the reason for their consideration as complex in-between atomic aggregation states) to extremely slowly segregating and seemingly permanently stable colloid mixtures, such as Faradey's gold sols that can still be observed in the British Museum. Complex interfaces of colloid systems are reminiscent to primitive cellular membranes, which in certain cases makes the overall environment wherein the reactions of synthesis take place sufficiently organized for complex self-organization processes (that are implicit characteristics of reactions in biomolecular domain) to occur. As a matter of fact, it has been proposed that inverse micelles formed by molecules of palmitic, stearic and oleic acids and glycerine at the air-water interface along the coastlines of some ancient seas presented primitive membranes that sustained the first planetary self-replication chemical reactions [172,173,174] as dynamic precursors for the evolution of life [175]. Colloid systems can provide complex interfaces that can support parallel reactions leading to surprisingly complex outcomes, and their relatively stable existence in thermodynamically metastable states can support significant changes in the products structures by the pure influence of aging treatment. However, small yields obtained due to employing extremely small concentrations and expensively complex environments used in most of the cases, altogether with the fact that increasing the space of options for production of various end-results via extremely fine variations of certain experimental parameters comes at the price of increased sensitivity of initial experimental settings that lead to reproducible outcomes, provides implicit difficulties in the advanced practical colloid science approach.

Apart from the use of reverse micelles in preparative organic chemistry for compartmentalization and selective solubilization and transfer of reactants, and separation of products [74], they have been used in the field of biochemistry both for storing bioactive chemical reagents [176] and as catalyzers [177,178] or inhibitors [179] of biochemical enzyme-driven reactions. Because encapsulating a protein in a reverse micelle and dissolving it in a low-viscosity solvent can lower the rotational correlation time of a protein and thereby provide a strategy for studying proteins in versatile environments [180], reverse micelles are used as a cell membrane-mimetic medium for the study of membrane interactions of bioactive peptides [181]. The observations that denaturation of proteins can be prevented in reverse micelles [182] have spurred even more interest in the application of these self-organized multi-molecular assemblies as either drug-delivery carriers or life-mimicking systems [183]. Such a biomimetic role of reverse micelles has been further instigated by the discovery of possibility of initiating self-replication of micelles and vesicles due to reactions occurring within their interiors. Vesicles and reverse micelles as separate 'microphases' and 'nanophases', respectively, may facilitate self-assembly and self-replication processes that would not be possible to occur in homogenous aqueous media. It was suggested that observations of micelles, vesicles and other self-assembled surfactant-composed structures as included in spontaneous formation of self-replicating order and selective compartmentalization of reactants ought to put these aggregates at a focal point for research into the origin of life [184]. As a matter of fact, positioning reverse micelles and other self-assembly colloid entities at the interface between the domains of 'living' and 'non-living' may present a crucial shift towards improved understanding of their 'templating' function and bioimitative utilization of such knowledge for practical purposes.

Both self-organization phenomena in living organisms and self-assembly effects of amphiphilic mesophases are governed by multiple weak interactions, such as hydrogen bonds, hydrophobic and hydrophilic interactions, van der Waals forces, salt bridges, coordination complexes (forces involving ions and ligands, i.e. 'coordinate-covalent bonds'), interactions among π-electrons of aromatic rings, chemisorption, surface tension, and gravity [185]. Whereas the traditional field of chemistry was developing by understanding the effects of covalent, ionic and metallic bonding forces, an extension of the same approach to weak intermolecular forces is nowadays suggested as a natural direction for achieving future prosperity within the practical aspect of the field of chemistry [186]. It should be noted that although first self-assembly, layer-by-layer techniques employed electrostatic forces (including organized networks of charged polymers), currently there is a trend of continual advancement in the development of self-assembly methods that rely on numerous other intermolecular interactions [187]. And it is the information (i.e. symmetry) stored within these weak bonding motifs that is in the latter cases 'coded' into the final self-assembled molecular architectures with an endless potential variety of symmetries depending on the complex interaction and synergic arrangement of various intertwined weak molecular forces [188]. Such molecular assemblies of unusual symmetries form spontaneously, sampling an ensemble of possible configurations and moving to a 'favorable' minimum free-energy state relative to many alternative and less 'favorable' structures that could form. Since non-covalent bonding with energies of relatively small multiples of kT dominates these processes, the ability to sample many possible configurations is implied as an inherent characteristic of such formed, typically highly organized macromolecular and multi-molecular spatial arrangements. As is well-known from protein folding experiments, extremely minor modifications of structure of precursor assemblers (corresponding to the primary structure of a protein, i.e. internal amino acid sequence) can lead to significant differences in the final structure (corresponding to the tertiary structure of a protein, i.e. conformational structure, non-covalent bridges, folded shape and exposition of active groups) and functional properties of the assembled objects. However, whereas on one hand large sensitivity of specific overall configurations and inability to control formation of final states emanate from this picture as necessary features of these, either biologically or synthetically processed systems, defect resistance on the other hand emerges as an apparently advantageous natural trait of self-assembling and spontaneously optimum configuration seeking systems [189]. Nevertheless, the most important point to be noticed here is that although fidelity and precision of information content (seen as internal guidance for spontaneous assembly) is essential for effective self-assembly of intricate, functional structures, biological systems do not synthesize perfectly [189]. Instead of humanly imagined 'perfect' design that never looks back and corrects mistakes, an essential element of natural synthesis pathways is detecting and correcting errors en route to the final product. With an eye towards biological processing routes, the concept of process-efficiency would therefore need to be revised and redefined if the science is about to achieve the control over molecular architecture that would resemble the one existing in the natural world.

Whereas on one hand elucidation and documentation of numerous synergies between weak interactions takes place in parallel with exciting discoveries of innovative applications thereof, on the other hand the greater employment of weak interactions in colloidal design necessarily means the greater frequency of imperfections, including both inability to form perfect lattices for as many repeat distances as those found in covalent, ionic or metallic materials [186] and inability to maintain the very reproducibility of internal structures once obtained. Therefore, as the aimed materials will be getting ever finer in their internal structure, a need to move from simply controlling global free energy minima to manipulation with the local ones would necessarily emerge as challenging, although the existence of a limit in both modeling attempts (that may come prior to fabrication) and processing control would eventually need to be acknowledged.

Nature is abundant with structures and processes that may serve as perfect models not only for metaphorical inspiration as a way of problem-solving, but for structural and processing imitation as well. Self-assembling mineral nucleation and growth in the nacreous layer of abalone sea shells, regulated by an organic template structure and resulting in a hybrid composite consisting of oriented columnar calcite crystals that provide strength and hardness to the shell and the inner lamellar composite of aragonite platelets and organic soft tissue, presents a widely studied example of biologically synthesized material [14]. Again, some animals can detect certain chemicals with sensitivities of only few molecules and thus provide excellent models for biomimetic development of advanced sensors with extreme sensitivity and selectivity [186]. Milk as the only food inherently produced by humans for humans, contains unique colloid structures, such as the colloidal calcium-protein complexes (casein micelles) or the milk fat globules with their unique membrane structures, and as such presents a perfect study case for the biomimicry application in the practical areas of colloid chemistry [190]. Egg yolk as not a simple protein solution, but a complicated dispersion of both micellar and granular lipoprotein particles, with proteins interacting strongly with surface-active phospholipids (lecithin) and neutral glycerides from the interior of the particles, and the stabilizing proteins that dominate the particle surfaces, may present another study example of an intricate biological colloid [191]. So far, mimicking simpler colloid formation mechanisms at the scale of laboratory synthesis have led to obtaining important novel structures. For example, mimicking the natural formation of opal - according to which the voids between ordered sediments of silica particles become infiltrated by hydrated silica which then solidifies - has already opened novel preparation pathways for a large number of templated mesoporous materials [144,192], whereas investigations of the complex relationships between numerous other biomineralized surfaces and the inherent organic matrices that are believed to attract ionic species and then control the subsequent, highly selective heterogeneous nucleation and growth of crystal surfaces (although there are theories on the origin of life that propose templating effects in the opposite direction [23], suggesting that mutually templating, feedback effects are present herein as well) continuously provide novel insights into applicable, self-assembly surface functionalization effects [193].

Certain characteristics of the processes of synthesis of such biological structures may become apparently evident. For instance, multifunctional biological materials are typically synthesized under conditions near room temperature, which favors self-assembly colloidal processes at mild heat contents and pressures as biomimicry ideals comparing to traditional and other high-energy methods of synthesis. Functional aspects of self-replication and self-repair phenomena present other attractive inherent features of natural self-assembly processes, whereby their utilization within synthetic self-assembly materials may be leading to novel, smart and dynamically self-maintainable structures [186]. Besides having delicate microstructures down to the atomic level, biological functional structures are also typical of being highly organized on all length scales, in a way that instead of apparent uniform aggregations at all scales, significant links exist between various levels of their organization. Biological structures are self-assembling, self-replicating and self-reproducing (so that the role of each substructural entity of any living system is the continual production of others), thermodynamically open but operationally closed systems wherein each cause presents an effect and *vice versa* [171,194]. Such an organization wherein circular causalities and feedback-pronounced regulation are emphasized instead of hierarchical disposal may prove to be a valuable ideal for a future bioimitating production of functional superstructures. However, as structural and organizational complexity of the designed mesoscopic systems approaches the circular complexity of biological systems, an increased sensitivity upon the smallest deviations from sets of standard processing and functional conditions - such as the one evident, for instance, in form of homeopathic or acupunctural effects in non-linear, thermodynamically open but operationally closed, autopoietic systems – as well as an increased challenge to control or modify the whole concept of control over the functional performance of such systems need to be expected.

There are many other questions related to practical colloid science that remain unanswered. For instance, the author's opinion is that too much attention is being paid today to superficially attractive systems, particularly with the advent of numerous microscopy techniques. Since there is no uniformity in natural reproduction (no two leaves or grains of sand are identical), maybe the ideals of the future would be not in producing perfectly uniform and identical particles, but in the production of unique structures, at the same time irreproducible and reliable upon application. In that sense, instead of emphasizing significance of substances with uniform arrangement of structural boundaries, materials with fractal [195] or fuzzy [196] internal patterns may prove to be a choice of the future. In other words, if Nature never produces absolutely identical entities in its design and self-replication processes (and we have learnt that mistakes in replication processes are key to the evolution of life), are then the tendencies towards perfect uniformities, symmetries and reproducibilities of today's fine structural design in true harmony with her? What would be the way to produce perfections through imperfections, as Nature does? On the other hand, as far as the current state in the field of materials synthesis is concerned, producing extraordinarily uniform systems indeed increases natural diversity of its structural components and thus presents novel and fascinating design achievements that result in perfectly reproducible properties and performances of the obtained materials, which is what society today mostly desires. However, as an array of seashore pebbles may be more fulfilling for contemplation than a necklace comprised of perfectly round and identical pearls, maybe the materials design of the future would increasingly be switching towards production of unique properties at all levels - from the constitutive particles to applicable devices that employ their functionalities - by the passage of time.

In comparison with the artificial crystals wherein perfect symmetries and uniformities are widely regarded as the carriers of performance fidelity, biotic crystals are typified by high surface roughness and intracrystalline protrusion of various subtle additives or impurities. Macromolecules, such as proteins, polysaccharides and lipids, are both key functional components of such biotic crystals (that improve mechanical and thermal response of the given materials, for example) and residues of mineral-organic matrix interactions inherent in their formation mechanism [197]. Also, despite the fact that biomineralization processes are based on using only few simple inorganic minerals such as $CaCO_3$, SiO_2 or hydroxyapatite, they can result in extremely large variety of materials properties [92]. The topography of enamel crystals exhibits surface irregularities of the order of size of single hydroxyapatite unit cells, which hypothetically corresponds to the tendencies to increase protein binding in the process of biomineralization [198]. Enamel crystal surfaces were shown to comprise a series of discrete and alternating domains of variously charged (in both magnitude and sign) surfaces, independently of topography, indicating periodical binding of a number of matrix proteins under physiological pH and ionic strength [198]. Such an 'imperfect' pattern of domains with various charges, surface protrusions and interleaving of individual crystal surfaces at the molecular level of biological materials stands in sharp contrast with the actual inclinations towards 'perfect' control of materials design at molecular scale among today's materials scientists and engineers.

In regard to inherent errors in the biological processes of genetic evolution, phylogenetic and ontogenetic learning, biomolecular catalysis, metabolic production and autopoiesis, the importance of the immanence of mistakes on many levels of scientific endeavor, generally ranging from the domain of investigation of correlations between processing settings and technological performance of materials to the domain of social presentation of research achievements, needs to be acknowledged as well. In that sense, capacity of the current peer-review methodology (that precedes manuscript publishing) to instigate research ideas and development directions ought to be seriously questioned, so that the degree of reliance upon the standard refereeing process may be conveniently reevaluated. Also, if assuming direct proportions between research quality and the number of its citations (or ISI impact factors of the publishing journals [199]) would be accepted as unreasonable as identifying musical quality with the number of sold records, new and more profound criteria for the estimation of research qualities need to be established. Realizing an unavoidable immanence of trial-and-error approach in materials design, every sort of result, either successful or not, would stand as of almost equal importance to a serious scientific mind, oriented not on a narrow target of immediate accomplishments, but on a larger Gestalt of natural order. For example, unsuccessful attempts of synthesis in specific compositional ranges of phase diagrams of certain colloid systems may mean a lot for the future understanding of the investigated areas of knowledge, though they are rarely shown within scientific presentations. Taking only single or few attractive points in investigated correlations between phase diagram positions and synthesized particle morphologies presents a routine 'resource-saving' procedure that ensures publishing success, although it presents an incomplete picture for the future understanding of the specific or even more generalized subject. However, the results of some of the most valuable scientific experiments in the history of science, such as Michelson-Morley and Lummer-Pringsheim experiments, Chadwick's investigation of β radioactive decay and Kepler's investigation of planetary motions, presented deviations from the actual explanation models and thus induced either their corrections or formation of new theories [200]. The novel trend of surpassing rigorous standard peer-review policy while ensuring an increased level of proliferation of breakthrough ideas or nuclei thereof may be noticed in the cases of preprint electronic server arXiv.org and online journal Philica. Whereas in the former case all articles are accepted for publishing without any form of refereeing, in the latter case a dynamic and open peer-review policy is used in a way that any researcher is invited to review an article after its unconditional publishing, with the comments being displayed at the end of each paper. The facts that the famous Watson and Crick's 1953 paper on the structure of DNA would hardly be published had the manuscript been peer-reviewed, and that Wikipedia, an open-source Internet encyclopedia, comprises approximately identical amount of information and proportion of errors as encyclopedia Britanica [201] may present a critical sign for final reevaluations of frequently too rigid and paradigmatic peer-review policies. Immanence of trial-and-error element within the colloid design practice implies that mistakes and 'not-how-to' present equally significant steps as attractive achievements and 'know-how' on the road towards overall advancement of the conceptual network of colloid chemistry. As an analogy, 'mind the gap' should always be the first thing to tell a child when stepping on a train, whereas the visions of final destinations may come later. Niels Bohr once observed that 'an expert is not a man who knows much about certain profession, but rather a man who has made all the mistakes, which can be made, in a very narrow field, and thus knows how to recognize, avoid and correct them'. And in a wide scheme of aspirations towards informational enrichment of human experiences and comprehension of nature and life, each small research, irrespective of its immediate or near- term successes, always stands as an enormously significant potential source of knowledge advancement and innovative achievements.

Chapter 7

MEETING OF THE VARIOUS DIRECTIONS OF FINE-STRUCTURAL DESIGN IN AN INTER-DISCIPLINARY NETWORK

Within each scientific research, theoretical and experimental efforts are much like left and right of a balanced tightrope-walker's walk. Whenever one begins to fall on either the side of overconceptualization without fruitful practical outcomes or overexperimentation without proper conceptual guidelines, one may know that the balance eventually needs to be restored. It is no doubt that significant achievements in theoretical correlations would mean that more predictable outcomes would come at a part of trial-and-error efforts, thus increasing the efficiency of practical approaches.

Whereas experimental efforts require a sort of conceptual framework for systematical summary of results and logical derivation of rules that will improve the predictability potential and increase efficiency of the experiments, many examples can support the idea that pure modeling achievements can anticipate experimental observations of certain phenomena. For instance, a recent model has shown that distribution of counterion atmosphere around isolated nucleic acid sequences is dependent on the slightest changes in initial positions of charged particles in the vicinity of nucleic acid [202], which is in accordance with the observations of the central role of non-linear self-organization effects in both colloid suspensions of DNA molecules [203] and cellular metabolic pathways [204]. Simulation approaches may thus obviously bring significant insights into the role that subtle deviations from perfectly periodic arrangement of both DNA strands and various biomolecular and polymeric configurations can play in defining correlation dynamics of closed loop formation, essential in explaining numerous reactions that involve interactions among macromolecular chains. On the other hand, an example from the classical materials science may invoke a recent proposal and experimental verification (hence, discovery) of a new mechanism of intermittent flow in small scale plasticity - based on the dislocations forming self-organized critical system and free surfaces limiting dislocation arm length for a distribution of sources - by means of dynamic dislocation simulations [205].

In fact, the whole set of achievements from the field of complexity theory predated actual employment of its concepts like fractal arrangement, bifurcations or butterfly effects in explanation of experimentally observed, natural phenomena [206,207,208]. As far as the colloid science is concerned, previous experience has demonstrated that even short simulations (nanoseconds on average) with highly simplified models of the constituents (bead-spring models of surfactants, for example) or pre-organized units (structurally constrained aggregates) may provide useful suggestions and creative stimulations in clarifying certain experimental notions [209]. Simulations may provide structural informations that are otherwise not accessible by means of experimental analysis, such as ultra-fine structural details within specific colloid aggregates. Novel self-assembled structures, their dynamics and phase transitions may be anticipated before experimental discoveries by means of adequate simulations, whereas acceptability of approximations in theoretical approaches (such as mean-field approximation is for example) and their underlying reliability can be systematically analyzed by means of simulation-guided experiments. Also, as possibly complementary to traditional, hard modeling approaches, soft modeling approaches, based on algorithms that look for certain patterns among sets of uncorrelated data, are being suggested as the next big breakthrough in the computational aspect of the modern chemistry.

Nevertheless, some of the questions and challenges for the simulation approaches in colloid science, related to their inherent limitations that stand on the way of the current drive to wed molecular dynamics simulations to experimental mesoscopic methods of particle design, may be mentioned at this point. First of all, with increasing the complexity of model settings and thus moving away from drastic simplifications that prevent a systematic examination of certain aspects of structural or temporal features of the system, simulation run times increasingly head into impractical ranges. Whereas the ranges of nanoseconds are typically involved in the contemporary simulations of colloidal effects, the time scale of interest may as well run into microseconds for mesoscale events or hundreds or more seconds for macroscale phenomena such as shear-induced changes [209]. Individual relaxation processes that determine dynamic membrane effects viewed as collective motion phenomena may span across time scales that differ by more than 10 orders of magnitude, such as smooth undulations of red blood cells with amplitudes in the range of micrometers and relative displacements of individual molecules that occur in the pico-second range and at lengths of less than the membrane thickness (~ 5 nm) [210]. Also, the standard description of bond cleavage processes (that can, in turn, be used to explore a potential energy landscape of binding) in theoretical modeling has so far been given by Kramers escape model that is formulated as a one variable problem (the variable being the bond length) and studied as a particle escape from a kinetic trap with a finite barrier under the influence of thermal noise and dissipation [211]. However – closely related to the fact that with an ongoing phase transition, the critical energy barrier is subject to change - whereas at the nanoscopic level a bond is indeed formed between two molecular entities, the participating entities are typically part of some extended structure, a membrane or a polymer, involving the existence of extra dynamic degrees of freedom to which the bond is coupled and resulting in multidimensional escape phenomenon that frequently may lead to unexpectedly rich phase behavior, dependent on subtle changes in initial settings of a model [212]. Then, in order to study colloidal templating effects, the structures of colloid aggregates, such as micelles, are *a priori* selected, which prevents a study of feedback effects related to mutual structural transformations between 'template' and a 'templated' material. Substructure and the internal degrees of freedom of the modeled particles are traditionally neglected, despite the fact that, although vital insights into phase behavior can be gathered, the properties and functions of certain molecular or multi-molecular aggregate structures, such as proteins, cannot be understand without invoking their atomic structure [213]. The laws of quantum mechanics, which can be useful for describing molecular geometry and energetics, have been so far completely disregarded in the field of computer simulation of surfactant systems [214], even though it is known that conformational transitions in polymers, including Levinthal paradox, can be properly explained by relying on the framework of quantum theory [215,216]. Also, hydrogen bonding in water as one of the fundamental biochemical interactions involved in enzyme catalysis, protein folding, DNA base-pairing, respiration and photosynthesis, comprises highly delocalized protons, indicating that quantum delocalization effects play central role in development and maintenance of life processes [217]. DNA base tautomerization as a form of proton tunneling, as well as the hypothesis according to which enzymes that establish the proton gradient that drives F1-ATPase use electron tunneling to connect the proton transport process to proteins of the respiratory chain may present additional examples of the subtle quantum mechanical basis of biochemistry [217]. However, we should be reminded that *ab initio* calculations that are frequently accepted as the basis of any comprehensive simulation approach, are based on approximate solutions of quantum mechanical problem settings (i.e. quantum mechanical energy terms are assumed as functional of local electron densities) and as such do not present 'first principles' calculations at all [129]. Finally, interactional molecular-level details, such as ion binding and chemical specificity, have often been ignored in order to simplify the simulation approaches and predict mesoscopic details, even though it is known that subtle effects, such as hydrogen bonding of co-surfactants with the head groups of surfactant, could play a key role in stabilizing colloid aggregates [209]. In essence, it is clear that mathematical modeling and related computer simulations are at present not sophisticated enough to produce precise quantitative predictions regarding behavior of colloid systems, and it has been almost generally acknowledged that the complexities encountered in most of the colloid systems regularly applied in various industrial fields (foods, pharmaceutics, cosmetics, etc.) is such that it is unlikely that this situation will change dramatically in the foreseeable future [218].

However, it is highly probable that the actual explanation mechanisms and principles used to describe self-assembly phenomena (such as selective attachments that constrain or favor certain crystal growth directions – which have never been directly observed during nanocrystal growth [127] - or other mechanistic representations, albeit the concepts of dynamic solvation according to which surfactants exchange on the nanocrystal surfaces during growth, and provide opportunities to more complex molecular recognition effects) will in the coming era increasingly cede their place to more abstract and hard to visualize models of colloid evolution and dynamic organization, that may more closely approach the real complexity of colloid systems. As the planetary model of the atom was replaced by the complex calculi of probability distributions in quantum theory (giving rise to many 'mystical' interpretations thereof, due to disappearance of a common sense in the visualizing representations of previously 'picturesque' describable phenomena [219]), and as feedback-based, non-linear relationships seem to pervade all biological phenomena, invoking systems theory that relies on an intricate mathematical framework in predicting or qualitatively approaching the observed phenomena, maybe with the advent of colloid science on both experimental and theoretical frontiers, current explanation paradigms will be abandoned on the account of unvisualizable abstract concepts, as such indeed leaving self-organization phenomena to the domain of purely practical and distantly wondering. Considering the actual growth in emphasizing non-local effects as quantum entanglement at atomic dimensions [220] and as 'butterfly' effects in feedback-permeated, biological and ecological systems [221,222], maybe the dynamics of microscopic aggregates, including the colloid systems with critical boundaries at micro-scale will not be immune to implying non-local fields and other non-mechanistic depictions within certain phenomenological explanation mechanisms for so long, whereas Sogami-Ise potential, non-local hypernetted-chain approximation or non-local density functional theory [223,224] may serve as neat examples for contemplation of the future progress in outlining the 'hidden connections' that might gradually emanate as the field of colloid science keeps on developing. In any case, in order to cover all length and time scales (from macroscopic to mesoscopic order) in describing and eventually predicting the behavior of non-equilibrium colloid systems by employing the conceptual framework of irreversible thermodynamics, based on closed set of non-linear equations of motions, the discrete nature of the constitutive particles cannot be ignored on mesoscopic scale [225]. However, similar to the living systems, non-local and memory effects would in that case become obviously decisive in determining key properties and the dynamic evolution of investigated colloid systems [225]. Also, due to a crucial role of interface effects in defining resulting colloid properties and functions, colloid systems cannot be represented as functions of the sums of their constituent phases. With a growth in colloidal complexity, their properties will become more and more resistant to simple additive models and model patches, thus clearly requiring non-additive, holistic approaches in understanding their behavior. Coping with or even instigating such an intricating shift in representation attitude would present an important challenge for the future simulations of colloid systems. And maybe the future era will witness new generations distantly recalling the times when old, coarse and primitive metaphors dominated the descriptions of phenomena under investigation, and thus, after ages of rationalism, maybe the age of new 'mysticism' in science - although hopefully pragmatic and even more rational in certain aspects (in acknowledging the immanence of uncertainties, impossibilities, the limits of knowledge and the fields of unknown, for instance) – is coming.

Nevertheless, the following example can illustrate the idea that certain modeling attempts can indeed serve as significant initiations for revision of current scientific representations, even at the most fundamental level. Recent attempt to model the structure of a reverse micelle has resulted in an image of a multi-molecular aggregate with surfactant head groups not completely shielding aqueous interior of the modeled micelle, leading to re-evaluation of the typical representations of micelles as perfectly surfactant-capped and too statically configured molecular aggregates [226]. Accordingly, neither lipid membranes can be seen anymore as passive matrices for hosting biomolecular reactions [227], confirming that cellular activities are in large extent controlled by lipids in addition to conventional protein-governed mechanisms [228]. Although it is known that chemical self-replication reactions need a sort of protection membrane to selectively absorb the influences of the environment, that is to say require "a sophisticated cradle to be lulled in" [229], how these protective mesophases indeed 'sing' presents a challenge for the future investigations. Knowing that by actively regulating the flow of chemicals between the cell and its surroundings and conducting electric impulses between nerve cells, biological membranes play a key role in cell metabolism and transmission of information within an organism highlights the practical significance of investigations oriented towards reproducing or at least approaching a reproduction of such an organizational complexity in artificial colloid systems. Also, knowing that malignant cells have significantly different surface properties comparing to normal cells [230], maybe the transition of focus in apoptosis researches away from the genetic code disruption as the sole key influence towards information transmission mechanisms that involve membrane mediation would herein beneficially switch the major scope to the cellular epigenetic network and finally to more holistic biological and biomedical perspectives. Such an integrative view at cellular structures may be further instigated by the recent findings that a large percentage of body cells (cardiac muscle cells, in particular) is, similar as the reverse micelle model [226], in a 'membrane-wounded' state, suggesting that continuous protective barrier is not essential for cell functioning [162]. If the cytoplasmatic medium is, instead as an ordinary solution, considered as a colloid gel, rich with interfaces between water and intracellular proteins, polysaccharides, nucleic acids and lipid membranes, then the water-retaining properties of cellular gel matrix may be reasonably accepted. Therefore, due to complex organization of the inherent interfacial areas, colloid systems may prove to be excellent models for providing significant links among both the fundamental fields of physical chemistry and biology, and their practical implementations in the areas of technological development, biomedicine and the science of sustainability.

It has been argued that enormous portion of the success in employing an array of modern chemical engineering methodologies came due to adoption and mastering of certain mathematical techniques that succeeded in keeping numerous growing peripheral fields during the second half of the 20th century around a strong and creative core of this practically oriented field [115]. However, in order to successfully overcome the approaching Schumpeter's crisis of rejuvenation with connecting to the innovative wave of introduction of ultra-sensitive, highly-organized, self-assembly systems in advanced technologies, a new set of modeling tools, related to complexity science apparatus ought to be adopted and implemented as practical platforms for chemical operation analysis and optimization. In this sense, a switch from standard, mechanistic and quantitatively predicting tools to qualitative, pattern-depicting models, such as via implementing the concepts of nonlinear dynamics, agent-based modeling, cellular automata, the aspects of game theory, statistical mechanics, graph and network theory, information theory and genetic algorithms needs to be stimulated within the framework of colloid science as well [112]. In general, systems engineering of the future would also need to face typical peculiarities of biological systems in modeling complex self-assembly pathways, such as inability to predict the behavior of the whole based on analysis of the elementary building blocks of the system ('superlattices' as certain self-assembled atomic aggregates have been shown to possess such holistic features [186,189]), inability to predict emergent qualities of self-organizing systems based on proposed quantitative relationships among its constituents (as is evidenced in examples ranging from qualities of consciousness to patterns of granular dynamics), inability to form comprehensive insight into spectrum of qualities of investigated systems with disregarding natural context of their existence, and the overall challenge to cope with error-tolerant, self-correcting, adapting and self-steering complex, either natural and synthetic systems where everything is indeed connected to everything else. No matter how molecular chaperones as certain 'catchers in the rye' of the cellular world (i.e. proteins that help other polypeptides to reach a proper conformation or cellular location without becoming part of the final structure) may look fascinating, someone has to build even them, and in the living and healthy autopoietic world, just like genes form proteins that form genes [171], everything is devoted to building everything else, from the cellular level to ethical and aesthetical social communities. Such a circular and holistic organization of the functional structural networks in the coming era will certainly present an enormous challenge that may require a paradigmatic shift in both applied theoretical tools and an overall attitude towards structural manipulation and materials design control.

Similar to the idea of inextricable complementarity between experimental and modeling approaches in colloid science, the research area of interest for understanding colloidal phenomena and using this knowledge for pragmatic purposes, may be divided into two parts [229]. On one side are analytic investigations of biomolecular recognition processes, driven by biocrystallography and other structural biology techniques and emphasizing local interaction features, conformational changes, stereo-specificity and unique molecular design as keys to understanding association effects on larger scales. On the other side is synthetic approach that is based on investigating large-scale self-association processes, relying on the concepts of weak intermolecular forces, entropy and statistical models of cooperative effects. Different levels of investigation in traditional fields of molecular biochemistry and soft matter - both relevant to the practical scope of colloid science - imply different explanation mechanisms as well, being exceedingly difficult to unify into a complete picture. All of the complex biochemical processes proceed via coupling conformationally specific binding on molecular scale with the cooperative tuning into higher biological contexts of their purpose by means of self-assembly phenomena that take place on typical colloid scales. As in the cases of confrontation between reductionistic and systemic methodologies of scientific analysis, the most significant pragmatic results will arise out of the compromise and mutually consistent application of both levels of scientific inquiry. In each case, since scientific models and physical qualities can more reasonably be viewed as human concepts with pragmatic purpose of mutual co-ordination of our experiences than as objective representations of universal reality [194,231], the significance of each scientific approach should be based on its pragmatic potential, that is value from the practical point of view. Nevertheless, it should be kept in mind that the most pragmatic scientific theories and discoveries have rarely achieved their practical meaning and significance immediately upon their announcement and affirmation. The practical significance of quantum chemistry, molecular biology, cybernetics or theory of relativity has been, for instance, continuously developing since these scientific fields and their paradigmatic hypotheses were proposed and recognized by scientific society.

In accordance with the proposed complementary approaches, weak soft-tech potentials for structuring self-assembled products into functional systems of hierarchical organization may be overcome by constructive coupling the soft-tech production of fine-structured materials with hard-tech assembly methodology. Excellent achievements have been recently reported by relying on such an approach [20,99,144,232,233]. Langmuir-Blodgett films [234], obtained by coupling self-assembled orientation of molecules at air-water interfaces with a technique for their deposition on solid substrates, present a classic example of such complementary methodology. Self-assembly parallelism and the selective patterning precision of lithographic and etching techniques can be united in a multitude of hybrid techniques for the production of fine structures [235]. In that sense, the prosperous lithographic field of microfluidics [236], coupled with parallel trial-and-error investigation of self-assembly patterns (currently known as microfluidic 'drug discovery') that may slightly differ in initial compositions of the precursor but lead to utterly diverse products, offers enormous potential for replicating the finest microfluidic and self-organization system known, i.e. human body. External fields, such as electric and magnetic fields, heat gradients or single layer shearing, can induce unexpected orderings apart from the usual long-range order and clustering in accordance with 'magic aggregation numbers', such as string-like ordering or triangular lattices depending on the intensities and directions of the field relative to the suspension cell [237]. On the other hand, electrospraying, electrocoalescence and other methods that involve various external fields, may be used for ink-jet spraying, fluid atomization, phase and particle separation, thus improving the functionalization control of self-assembled fine structures [238]. However, whereas the so-called 'fat and sticky finger' problem, related to limited capabilities in precise manipulation of sufficiently small atomic or molecular units, presents the major problem of all proposed 'molecular assembly' techniques (including lithography) in the attempts to scale down the 'high-tech' design of ultra-fine supramolecular architectures, the fact that in most cases only periodic or quasiperiodic structures are obtained by self-assembling methods remains the major limitation of self-assembly approaches [239]. Nevertheless, an advance in fine control of local phase transitions within colloidal precursors presents a natural direction for the future progress along the line of such compromise between hard-tech manipulation and self-assembly processing.

The lithographic techniques for manipulation of matter have been one of the driving forces behind the miniaturization of modern functional devices, and the main problems currently associated with them in attempts to disprove expectations of Moore's law, such as inherent limits in attainable resolution and exorbitant cost of production tools, could be elegantly overcome by application of various self-assembling molecular structures [240]. Graphoepitaxy as a process wherein grooves with typically micron or submicron dimensions are patterned on a substrate by photolithography and etching, whereby the domain structure of block copolymer films deposited in the grooves nucleates on the walls of the topographic features and propagates inward so as to form well ordered and functional assemblies below the achievable resolution size for a given lithographic technique, presents an excellent example of combining so-called 'top-down' and 'bottom-up' fabrication approaches in synergetic processing units [241]. Thoroughly opposite strategy may also be employed; namely, precursor self-assembly phase could be deposited in the first step, directed into desired patterns by means of a lithographic technique or an external field in the second stage, after which chemical reaction and nucleation of final particles may be initiated. In that sense, electron-beam lithography had been successfully applied for immobilization of deposited diblock copolymer micelles and their crosslinking, which after hydrogen gas plasma treatment resulted in patterns of various geometries made out of uniform-sized gold particles [242]. Depending on numerous processing variables, such as chemical surface patterns (i.e. substrate topography and the relation between the typical periods of self-assembled molecular forms and surface pattern period), thermal annealing (i.e. eutectic solidification, crystallization, solvent evaporation, etc.), constituent polymer identities, number of monomeric units per copolymer block, molar ratio of a block copolymer to its corresponding homopolymers in the applied ternary blends, the degree of polymerization, Flory-Huggins interaction parameter, surface charging, deposited film thickness, external fields and shearing effects, various self-assembled structural patterns, scalable down to sub-10 nm dimensions, may be obtained. By sequentially or simultaneously applying multiple alignment effects induced by each one of these variables, numerous synergetic outcomes with exceptional long-range order, prepared in much shorter periods of time, have been achieved [240]. With using such conjunction of block copolymer self-assembly patterning and lithographic techniques, complex geometries such as the ones that are present in typical integrated circuits and memory arrays (e.g. bent liens with sharp corners, lines that end at specific positions, lines of varying width, T-junctions, jogs, arrays of spots, etc.) have recently been fabricated [241]. Unlike conventional lithographic techniques alone, their coupling with the use of self-assembled structures of block copolymer materials may result not only in finer structural geometries obtainable with improved resolution, but in achieving three-dimensional long-range ordering in single processing steps [241]. Also, comparing to traditional, diffusion-limited lithographic processing methods wherein small deviations in either the exposure dose or post-exposure annealing conditions can lead to large variations in the final structural features, thermodynamically controlled self-assembly processes would be able to correct of self-heal various irregularities, such as dimensional variations of defects in the chemical surface patterns. Such a presented synergetic coupling has thus clearly been demonstrated as a viable method for future fabrication of advanced microprocessors and integrated circuits.

Processing methods based on simultaneous combination of finely controlled deposition and self-assembly re-arrangement offer numerous options for production of diverse geometric orders and well-defined morphologies. For example, depending on the ratio between diffusion rate D and deposition flux F, subtle interplay between kinetic and thermodynamic effects can be set, reflecting on various potentially obtainable structures [99]. At large values of D/F, when deposition is slower than diffusion, adsorbed molecules have enough time to explore the potential energy surface and find the minimum energy configurations. Self-assembly processes of flexible molecular nanoscale patterns that rely on relatively weak, non-covalent interactions are obviously favored in such set of experimental conditions. Moreover, to allow for supramolecular assembly based on molecular recognition effects, conditions close to equilibrium are required, and such processes of post-deposition self-assembly are clearly governed by thermodynamics of the system. On the other side, at small values of D/F, when deposition is fast relative to diffusion, the pattern of growth is basically determined by kinetics, and metastable states become achievable, which is related to possibilities of manipulating guidance of molecular and atomic species into predetermined patterns. Because metal bonds do not possess directionality that can be used for directing interatomic interactions, and shapes and sizes of metallic clusters on the deposition substrates are mostly determined by available atom movements, such as diffusion on surface terraces, over steps, along edges and across corners or kinks, small D/F values provide the best conditions for fine structural arrangement of materials with such rigid structural features into mesoscopic structures of various symmetries, ranging from compact uniform clusters and large faceted islands to fractals, dendrites and atomically thin chains [99]. However, the middle ground where thermodynamics and kinetics (intermediate D/F values) become intermingled into a complex interplay presents potentially the most fruitful design space with an enormous variety of processing settings that may result in remarkably organized surface structures. However, this range offers numerous sensitive processing conditions, whereas so far incompletely understood post-deposition morphological evolution of quantum dots (due to the complex interplay between discrete energy spectra of the constitutive electrons and holes, and their structurally "tunable", solid-state features), typically prepared in intermediate D/F range, may support this fact [99]. In fine self-organizing resonance with additional external effects, such as shear, heat, composition and qualities of environment, electromagnetic fields, effects that will enable finely focused, spatial and sequential deposition control, and precisely set surface modulations in either periodicity of the supporting crystal structure, surface steps and terraces (as in the step-decoration method [243]) or certain controlled pulse-effects, such as strain-relief, further refinement in the produced mesoscopic order could be achieved. However, in order to apply so, a certain advance in understanding the influence of substrate atomic lattice and electronic structure ought to be initiated. For example, the applied substrates can frequently alter the structure of functional moieties or electronic structure of adsorbed ligands and *vice versa*, resulting in unpredictable final self-assembly outcomes, and making the solution-based coordination chemistry concepts impossible to apply without significant modifications [99]. Transformation of structurally sensitive nanoscale architectures into more rigid, functional structures while retaining their precise spatial organization, presents one more challenge that will gradually prove to be ever more significant as the complexity of synthesized supramolecular patterns keeps on increasing.

To sum up, leaving aside half-pieces of complementary wholes and unilaterally adhering to certain rigid creative directives may therefore not be the most fruitful approach in the propagation of any sort of creative efforts. But looking for the best combinations from diverse creative options and directions around us may instead present a correct attitude towards advancing all levels of science and knowledge. And in order for our creative aspirations to stretch their hands onto many crossways within the conceptual schemes of science, cultivating inter-disciplinary thirst, i.e. enormous inquiry that breaks the boundaries of specific fields and links them into novel, trans-disciplinary networks, presents surely an educational path for the technologically and humanly fruitful future.

Chapter 8

CONCLUSION

Let us at the end get back to where we started from, that is to the question whether perfect design of whatever functional structures imagined would be made possible in the future. By invoking different points of view on the same subject in each of the sections, wider and clearer perspectives could have been formed. However, it looks as if colloid science is still facing only horizons ending with many unknown and unforeseen options, possibilities and choices as far as our sight can reach. Although the trend of producing ever more complex, highly organized and intricately symmetric mesoscopic structures can be expected to continue, natural order on fundamental scale may as well not allow for formation of whatever fine structures imagined by visionary minds. As we have seen so far, the procedures for preparation of novel fine structures of controlled properties by using the practical methodology of colloid chemistry are not based on controlled manipulation of variables that figure in any of the fundamental natural laws. Instead, the practical chemists are limited to setting the right boundary conditions of experiments through trial-and-error manipulation of the given macroscopic scale parameters and letting Nature herself to perform the inherent physical transformations, after which the products may be collected and further used. That is to say, the modern design of fine material structures never 'touches' Nature on its fundaments, but only manipulates with the tracks upon which her trains are already heading. Every process of product and technology design is therefore self-organizing in its essence. Whereas questioning 'how' from this point on would lead the practical field of colloid science towards future production of novel and ever more attractive fine material structures, questioning 'why' would lead to opening of frequently forgotten but equally essential, philosophical, ethical and aesthetical stances of the chemical science.

Whereas molecular machining promises for perfect design of ultra-fine structures were challenged on many frontiers, the weaknesses of colloidal, self-organizing approach to design of novel materials and templating fabrication of functional structures were also outlined. Complex, difficult to define and control relationships between macroscale processing parameters and inherent system variables present the major reason for limiting the achievements of practical colloid science in large extent to the consequences of trial-and-error approaches. Moreover, we have realized that by further development in the complexity of colloid systems that practical colloid science will utilize (corresponding to their approach to higher complexity organization that is characteristic of living systems), a trend of increased sensitivity of the final product states upon initial conditions of the preparation processes and the overall history of the system may be expected. This will in certain extent lead to functionally more versatile and multifunctionally applicable structures, but in large extent it will at the same time follow a trend of limiting potential reproducibilities and precise control over final product properties below certain level of fine internal structuring. As with the other areas of practical science where artificial products are increasingly approaching the complexity of living systems (i.e. artificial intelligence, neural networks, molecular machining), the issues of increasing difficulties to control, desirably tune and reproduce will slowly appear as immanent and necessary to tackle. Colloid chemists have always been facing cybernetic 'black boxes' in describing the multitude of relationships between manipulated macroscopic parameters and internal variables of the systems in question. But thinking about how far the design in colloid chemistry will go before facing 'black boxes' in the very complexity of the final products and their response to corresponding stimuli may bring us to face more of the amusing horizons of the herein discoursed topic. Anyhow, the search for the common grounds between our design blueprints and the way of Nature, as well as between hard-tech manipulative and soft-tech self-assembling synthesis pathways emanates as a necessary directive from the herein presented perspective. Such an approach may lightly, step by step, lead us to outstanding features of 'technologies with the human face' [244,245] that will foster humanity and benefit the spirit as much as they will contribute to comfort and abundance.

As a conclusion, if perfect design would be possible, it would not be related to perfect copying of tectonics of our dreams to the substrate of reality, but to sensitively building the informational landscape of the world with careful and wise treading the wondrous embroidery of natural self-organization patterns. The balance of 'the freedom and the guide', i.e. invoking both intuitive, artistic and endlessly inspiring search and reliance on predictable guiding schemes within the design processes of colloid science, is therefore here to stay, and although it may look irritating at first, the natural gift of this balance may fill many a books with the praises thereof.

Alexander Pope's maxim that "where fools rush in, angels fear to tread" may remind us of a humble attitude that future colloid scientists ought to keep. It may provide us with the right balance of 'the freedom and the guide', and with the senses of uncertainty, questioning and wondering as genuine initiators of true scientific advancements. There will be no perfect models that might account for every particular colloid system in nature and enable perfect predictability of envisaged design procedures, at least not in the foreseeable future. Every individual system will keep on being an inexhaustible source of potential novelties and practical significance, and facing the challenges of colloid design will in the most promising scope continue requiring subtle combinations of general knowledge and specific originalities introduced.

Finally, in the endless search for the perfect design and perfect models to describe the world, we should occasionally ask a fundamental question: would we really want to be in possession of perfect design abilities? Maybe the ethics of human race is still far from being able to sustainably cope with the consequences of perfect design, so far requiring assistance from self-organization patterns of Nature. In this case, as senses of responsibility and deep ethics are being developed, natural secrets may unravel in parallel and through scientific pursuit keep on providing us with ever more refined models for design. Or on the other hand, it may be that continual quests present necessary prerequisites for the development of human ethics and other divine qualities, leaving us to contemplate on the seemingly paradoxical, but in fact marvelous dynamic balance between vigorous quests for knowledge and facing perpetual, although ever brighter horizons thereof.

And in the end, let us be reminded how balancing between alternate and complementary sides of the mentioned tightrope-walker's walk presents necessary precondition for his advancing along the line. In a similar way, only through constant searches for restoring balances among various complementary sides presented herein (as some of the key branches in the line of the future development of the discoursed fields) is that both designer and society satisfaction and natural diversity will flourish in a sustainable and harmonic way.

REFERENCES

[1] P. G. de Gennes – "Ultradivided Matter", *Nature* **412** (6845) 385 (2001).

[2] M. Faraday – "Experimental Relations of Gold (and Other Metals) to Light", *Philosophical Transactions of the Royal Society* **147** (1), 145 – 181 (1857).

[3] P. J. Wilde – "Interfaces: Their Role in Foam and Emulsion Behaviour", *Current Opinion in Colloid and Interface Science* **5**, 176 – 181 (2000).

[4] J. Sjöblom, R. Lindberg, S. E. Friberg – "Microemulsions – Phase Equilibria Characterization, Structures, Applications and Chemical Reactions", *Advances in Colloid and Interface Science* **95**, 125 – 287 (1996).

[5] V. Uskoković, M. Drofenik – "Synthesis of Materials within Reverse Micelles", *Surface Review and Letters* **12** (2) 239 – 277 (2005).

[6] S. E. Friberg – "Microemulsions", *Progress in Colloid & Polymer Science* **68**, 41 – 47 (1983).

[7] J. H. Schulman, W. Stoeckenius, L. M. Prince – "Mechanism of Formation and Structure of Microemulsions by Electron Microscopy", *Journal of Physical Chemistry* **63,** 1677 – 80 (1959).

[8] T. Hellweg – "Phase Structures of Microemulsions", *Current Opinion in Colloid and Interface Science* **7**, 50 – 56 (2002).

[9] V. Uskoković, M. Drofenik – "Synthesis of Nanocrystalline Nickel-Zinc Ferrites within Reverse Micelles", *Materials and Technology* **37** (3-4) 129 – 131 (2003).

[10] V. T. John, B. Simmons, G. L. McPherson, A. Bose – "Recent Developments in Materials Synthesis in Surfactant Systems", *Current Opinion in Colloid & Interface Science* **7**, 288 – 295 (2002).

[11] M. Gradzielski – "Investigations of the Dynamics of Morphological Transitions in Amphiphilic Systems", *Current Opinion in Colloid and Interface Science* **9**, 256 – 263 (2004).

[12] J. Patarin, B. Lebeau, R. Zana – "Recent Advances in the Formation Mechanisms of Organized Mesoporous Materials", *Current Opinion in Colloid and Interface Science* **7**, 107 – 115 (2002).

[13] "Molecular Self-Assembly: An Interview with Ralph G. Nuzzo, Ph.D.", *Essential Science Indicators Special Topics*, retrieved from http://www.esi-topics.com/msa/interviews/RalphNuzzo.html (August 2002).

[14] J. Liu, A. Y. Kim. L. Q. Wang, B. J. Palmer, Y. L. Chen, P. Bruinsma, B. C. Bunker, G. J. Exarhos, G. L. Graff, P. C. Rieke, G. E. Fryxell, J. W. Virder, B. J. Tarasevich, L. A. Chick – "Self-Assembly in the Synthesis of Ceramic Materials and Composites", *Advances in Colloid and Interface Science* **69**, 131 – 180 (1996).

[15] V. Uskoković – "Nanotechnologies: What We Don't Know", *Technology in Society* **29** (1) 43 – 61 (2007).

[16] R. Feynman – "There's a Plenty of Room at the Bottom: An Invitation to Enter a New Field of Physics", Lecture at the California Institute of Technology (December 29, 1959).

[17] K. E. Drexler – "Engines of Creation: the Coming Era of Nanotechnology", Anchor, NY (1986).

[18] M. V. Tirrell, A. Katz (eds.) – "Self-Assembly in Materials Synthesis", *MRS Bulletin* **30** (2005).

[19] Z. Hassan, C. H. Lai (eds.) – "Ideals and Realities: Selected Essays of Abdus Salam", World Scientific, Singapore (1984).

[20] T. Kraus, L. Malaquin, E. Delamarche, H. Schmid, N. D. Spencer, H. Wolf – "Closing the Gap Between Self-Assembly and Microsystems using Self-Assembly, Transfer, and Integration of Particles", *Advanced Materials* **17** (20) 2438 – 2442 (2005).

[21] R. Baum – "Nanotechnology: Drexler and Smalley Make the Case For and Against 'Molecular Assemblers'", *Chemical & Engineering News* 81 (48) (December 1, 2003).

[22] R. E. Smalley – "Of Chemistry, Love and Nanobots", *Scientific American* **285** (3) 68 – 69 (2001).

[23] L. L. Hench – "Science, Faith and Ethics", Imperial College Press, London (2001).

[24] L. Glass – "Synchronization and Rhythmic Processes in Physiology", *Nature* **410**, 277 – 284 (2005).

[25] G. Bateson – "Mind and Nature: A Necessary Unity", Hampton Press, Cresskill, NJ (2002).

[26] P. R. Ehrlich, A. H. Ehrlich, J. P. Holdren – "Ecoscience: Population, Resources, Environment", W. H. Freeman and Co., San Francisco, CA (1977).

[27] T. Kuhn – "The Structure of Scientific Revolutions", Routledge, New York (2002).

[28] K. Popper – "The Logic of Scientific Discovery", Routledge, New York (2004).

[29] N. Aubry, P. Singh – "Control of Electrostatic Particle – Particle Interactions in Dielectrophoresis", *Europhysics Letters* **74** (4) 623 – 629 (2006).

[30] K. Wormuth – "Superparamagnetic Latex via Inverse Emulsion Polymerization", *Journal of Colloid and Interface Science* **241**, 366 – 377 (2001).

[31] N. S. Kommareddi, M. Tata, V. T. John, G. L. McPherson, M. F. Herman, Y. S. Lee, C. J. O'Connor, J. A. Akkara, D. L. Kaplan – "Synthesis of Superparamagnetic Polymer – Ferrite Composites Using Surfactant Microstructures", *Chemistry of Materials* **8**, 801 – 809 (1996).

[32] P. Tartaj, C. J. Serna – "Microemulsion-Assisted Synthesis of Tunable Superparamagnetic Composites", *Chemistry of Materials* **14**, 4396 – 402 (2002).

[33] D. G. Grier – "A Revolution in Optical Manipulation", *Nature* **424**, 810 – 816 (2003).

[34] P. T. Korda, G. C. Spalding, E. R. Dufresne, D. G. Grier – "Nanofabrication with Holographic Optical Tweezers", *Review of Scientific Instruments* **73**, 1956-1957 (2002).

[35] J. D. Moroz, P. Nelson, R. Bar-Ziv, E. Moses – "Spontaneous Expulsion of Giant Lipid Vesicles Indued by Laser Tweezers", *Physical Review Letters* **78** (2) 386 – 389 (1997).

[36] C. H. Chiou, Y. Y. Huang, M. H. Chiang, H. H. Lee, G. B, Lee – "New Magnetic Tweezers for Investigation of the Mechanical Properties of Single DNA Molecules", *Nanotechnology* **17** (5) 1217 – 1224 (2006).

[37] M. A. Brown, E. J. Staples – "Characterisation of the Weak Interactions Between a Particle and a Plane Surface Using Total Internal Reflection Microscopy and Radiation Pressure Forces", *Faradey Discussions of the Chemical Society* **90**, 193 – 208 (1990).

[38] J. J. Hawkes, J. J. Cefai, D. A. Barrow, W. T. Coakley, L. G. Briarty – "Ultrasonic Manipulation of Particles in Microgravity", *Journal of Physics D* **31**, 1673 – 1680 (1998).

[39] G. Myszkiewicz, J. Hohlfeld, A. J. Toonen, A. F. Van Etteger, O. I. Shklyarevskii, W. L. Meerts, T. Rasin, E. Jurdik – "Laser Manipulation of Iron for Nanofabrication", *Applied Physics Letters* **85** (17) 3842 – 3844 (2004).

[40] F. Rubio-Serra, H. Javier, M. Wolfgang, W. Robert – "Nanomanipulation by Atomic Force Microscopy", *Advanced Engineering Materials* **7** (4) 193 – 196 (2005).

[41] D. A. Tomalia, Z. G. Wang, M. Tirrell – "Experimental Self-Assembly: The Many Facets of Self-Assembly", *Current Opinion in Colloid and Interface Science* **4** (1), 3 – 5 (1999).

[42] R. F. Service – "Making Devices Smaller, Brighter, and More Bendy", *Science* **282**: 2179 – 2180 (1998).

[43] "Is the Revolution Real? Debating the Future of Nanotechnology", Foresight Institute Commentary and FAQ; http://foresight.org/NanoRev/istherev.html.

[44] G. Whitesides, B. Grzybowski – "Self-Assembly at All Scales", *Science* **295**, 2418 – 2421 (2002).

[45] J. I. Vontz, V. Vesna – "In the Future, Every Molecule Will Have Its 15 Minutes of Fame", *Los Angeles Times* (August 31, 2003).

[46] R. Jones – "The Future of Nanotechnology", *Physics World* (August 2004).

[47] V. Derjaguin and L. Landau – "Theory of the Stability of Strongly Charged Lyophobic Sols and of the Adhesion of Strongly Charged Particles in Solution of Electrolytes", *Acta Physicochimica (URSS)* **14** (6) 633 – 662 (1941).

[48] E. J. W. Verwey, J. T. G. Overbeek, K. Van Ness (Eds.) – "Theory of the Stability of Lyophobic Colloids – The Interactions of Soil Particles Having an Electrical Double Layer", Elsevier, Amsterdam (1948).

[49] Blume, Th. Zemb – "Self-Assembly: Weak Molecular Forces at Work for Building Mesoscopic Architectures", *Current Opinion in Colloid and Interface Science* **7**, 66 – 68 (2002).

[50] J. C. Crocker – "Interactions and Dynamics in Charge-Stabilized Colloid", *MRS Bulletin* **23**, 24 – 31 (1998).

[51] R. Coalson, T. L. Beck – "Numerical Methods for Solving Poisson and Poisson-Boltzmann Type Equations", *Encyclopedia of Computational Chemistry* **3**, edited by P. Von Rague Schleyer, pp. 2086 – 2100, John Wiley, New York (1998).

[52] J. A. Weiss, A. E. Larsen, D. G. Grier – "Interactions, Dynamics, and Elasticity in Charge-Stabilized Colloidal Crystals", *Journal of Chemical Physics* **109** (19) 8659 – 8666 (1998).

[53] M. Janosi, T. Tel, D. E. Wolf, J. A. C. Gallas – "Chaotic Particle Dynamics in Viscous Flows: The Three-Particles Stokeslet Problem", *Physical Review E* **56**, 2858 – 2868 (1997).

[54] Faybishenko, A. J. Babchin, A. L. Frenkel, D. Halpern, G. I. Sivashinsky – "A Model of Chaotic Evolution of an Ultrathin Liquid Film Flowing Down an Inclined Plane", *Colloids and Surfaces A: Physicochemical and Engineering Aspects* **192** (1-3), 377 – 385 (2001).

[55] Sogami – "Effective Potential Between Charged Spherical Particles in Dilute Suspension", *Physics Letters* **96A** (4), 199 – 203 (1983).

[56] G. Grier – "When Like-Charges Attract: Interactions and Dynamics in Charge-Stabilized Colloidal Suspensions", *Journal of Physics: Condensed Matter* **12**, A85-A94 (2000).

[57] G. Grier, S. H. Behrens – "Interactions in Colloidal Suspensions: Electrostatics, Hydrodynamics and Their Interplay", in *Electrostatic Effects in Biophysics and Soft Matter*, edited by C. Holm, P. Kekicheff and R. Podgornik, Kluwer, Dordrecht (2001).

[58] K. Zahn, G. Maret – "Two-Dimensional Colloidal Structures Responsive to External Fields", *Current Opinion in Colloid & Interface Science* **4**, 60 – 65 (1999).

[59] P. Attard – "Recent Advances in the Electric Double Layer in Colloid Science", *Current Opinion in Colloid and Interface Science* **6**, 366 – 371 (2001).

[60] W. Ninham – "On Progress in Forces Since the DLVO Theory", *Advances in Colloid and Interface Science* **83**, 1 – 17 (1999).

[61] H. K. Christenson, P. M. Claesson – "Direct Measurements of the Forces Between Hydrophobic Surfaces in Water", *Advances in Colloid and Interface Science* **91**, 391 – 436 (2001).

[62] G. Grier – "Colloids: A Surprisingly Attractive Couple", *Nature* **393** (6686) 621 (1998).

[63] M. Corti, T. Zemb – "Self-Assembly Under the Influence of Weak or Long-Range Forces", *Current Opinion in Colloid and Interface Science* **5,** 1 – 4 (2000).

[64] V. M. Gun'ko, V. V. Turov, V. M. Bogatyrev, V. I. Zarko, R. Leboda, E. V. Goncharuk, A. A. Novza, A. V. Turov, A. A. Chuiko – "Unusual Properties of Water at Hydrophilic/Hydrophobic Interfaces", *Advances in Colloid and Interface Science* **118**, 125 – 172 (2005).

[65] A. Vogler – "Structure and Reactivity of Water at Biomaterial Surfaces", *Advances in Colloid and Interface Science* **74**, 69 – 117 (1998).

[66] Bognolo – "The Use of Surface-Active Agents in the Preparation and Assembly of Quantum-Sized Nanoparticles", *Advances in Colloid and Interface Science* **106**, 169 – 181 (2003).

[67] Y. Chevalier – "New Surfactants: New Chemical Functions and Molecular Architectures", *Current Opinion in Colloid & Interface Science* **7,** 3 – 11 (2002).

[68] R. P. Bagwe, K. C. Khilar – "Effects of Intermicellar Exchange Rate on the Formation of Silver Nanoparticles in Reverse Microemulsions of AOT", *Langmuir* **16**, 905 – 10 (2000).

[69] U. Natarajan, K. Handique, A. Mehra, J. R. Bellare, K. C. Khilar – "Ultrafine Metal Particle Formation in Reverse Micellar Systems: Effects of Intermicellar Exchange on the Formation of Particles", *Langmuir* **12**, 2670 – 2678 (1996).

[70] N. E. Levinger – "Ultrafast Dynamics in Reverse Micelles, Microemulsions, and Vesicles", *Current Opinion in Colloid and Interface Science* **5**, 118 – 124 (2000).

[71] L. S. Romsted, C. A. Bunton, J. Yao – "Micellar Catalysis, a Useful Misnomer", *Current Opinion in Colloid & Interface Science* **2**, 622 – 628 (1997).

[72] K. Inouye, R. Endo, Y. Otsuka, K. Miyashiro, K. Kaneko, T. Ishikawa – "Oxygenation of Ferrous-Ions in Reversed Micelle and Reversed Micro-Emulsion", *Journal of Physical Chemistry* **86** (8), 1465 – 69 (1982).

[73] Li, G. Z. Li, H. Q. Wang, Q. J. Xue – "Studies on Cetyltrimethylammonium Bromide (CTAB) Micellar Solution and CTAB Reversed Microemulsion by ESR and ^{2}H NMR", *Colloids and Surfaces A: Physochemical and Engineering Aspects* **127**, 89 – 96 (1997).

[74] M. A. Lopez-Quintela, C. Tojo, M. C. Blanco, L. Garcia Rio, J. R. Leis – "Microemulsion Dynamics and Reactions in Microemulsions", *Current Opinion in Colloid and Interface Science* **9**, 264 – 78 (2004).

[75] S. Otto, J. B. F. N. Engberts, J. C. T. Kwak – "Million-Fold Acceleration of a Diels-Alder Reaction Due to Combined Lewis Acid and Micellar Catalysis in Water", *Journal of the American Chemical Society* **120**, 9517 – 9525 (1998).

[76] K. K. Ghosh, L. K. Tiwary – "Microemulsions as Reaction Media for a Hydrolysis Reaction", *Journal of Dispersion Science and Technology* **22** (4) 343 – 8 (2001).

[77] Z. Shervani, Y. Ikushima – "Compartmentalization of Reactants in Supercritical Fluid Micelles", *Journal of New Chemistry* **26**, 1257 – 60 (2002).

[78] E. Carpenter, C. T. Seip, C. J. O'Connor – "Magnetism of Nanophase Metal and Metal Alloy Particles Formed in Ordered Phases", *Journal of Applied Physics* **85** (8) 5184 – 5186 (1999).

[79] M. P. Pileni, T. Zemb, C. Petit – "Solubilization by Reverse Micelles: Solute Localization and Structure Perturbation", *Chemical Physics Letters* **118** (4), 414 – 420 (1985).

[80] M. P. Pileni – "Reverse Micelles as Microreactors", *Journal of Physical Chemistry* **97**, 6961 – 6973, (1993).

[81] P. A. Dresco, V. S. Zaitsev, R. J. Gambino, B. Chu – "Preparation and Properties of Magnetite and Polymer Magnetite Nanoparticles", *Langmuir* **15**, 1945 – 1951 (1999).

[82] P. Barnickel, A. Wokaun, W. Sager, H. F. Eickel – "Size Tailoring of Silver Colloids by Reduction in W/O Microemulsions", *Journal of Colloid and Interface Science* **148** (1), 80 – 90 (1992).

[83] J. Arriagada, K. Osseo-Asare – "Synthesis of Nanosize Silica in a Nonionic Water-in-Oil Microemulsion: Effects of the Water/Surfactant Molar Ratio and Ammonia Concentration", *Journal of Colloid and Interface Science* **211**, 210 – 220 (1999).

[84] C. C. Wang, D. H. Chen, T. C. Huang - "Synthesis of Palladium Nanoparticles in Water-in-Oil Microemulsions", *Colloids and Surfaces A: Physicochemical and Engineering Aspects* **189**, 145 – 154 (2001).

[85] R. Hua, C. Zang, C. Shao, D. Xie, C. Shi – "Synthesis of Barium Fluoride Nanoparticles from Microemulsion", *Nanotechnology* **14**, 588 – 591 (2003).

[86] D. O. Yener, H. Giesche – "Synthesis of Pure and Manganese-, Nickel-, and Zinc-Doped Ferrite Particles in Water-in-Oil Microemulsions", *Journal of the American Ceramic Society* **84** (9) 1987 - 1995 (2001).

[87] D. Roux, A. M. Bellocq, P. Bothorel – "Effect of the Molecular Structure of Components on Micellar Interactions in Microemulsions", *Progress in Colloid & Polymer Science* **69**, 1 - 11 (1984).

[88] J. C. Lin, J. T. Dipre, M. Z. Yates – "Microemulsion-Directed Synthesis of Molecular Sieve Fibers", *Chemistry of Materials* **15**, 2764 – 2773 (2003).

[89] R. Hua, C. Zang, C. Shao, D. Xie, C. Shi – "Synthesis of Barium Fluoride Nanoparticles from Microemulsion", *Nanotechnology* **14**, 588 – 591 (2003).

[90] J. D. Hines – "Theoretical Aspects of Micellisation in Surfactant Mixtures", *Current Opinion in Colloid and Interface Science* **6**, 350 – 356 (2001).

[91] G. Döbereiner, E. Evans, M. Kraus, U. Seifert, M. Wortis – “Mapping Vesicle Shapes into the Phase Diagram: A Comparison of Experiment and Theory”, *Physical review E* **55** (4) 4458 – 4474 (1997).

[92] W. Meier – “Nanostructure Synthesis Using Surfactants and Copolymers”, *Current Opinion in Colloid & Interface Science* **4**, 6 – 14 (1999).

[93] D. Chandler – “Interfaces and the Driving Force of Hydrophobic Assembly”, *Nature* **437**, 640 – 647 (2005).

[94] B. Lindman, H. Wennerstrom – “Micelles. Amphiphile Aggregation in Aqueous Solution”, *Topics in Current Chemistry* **87**, Springer-Verlag, Berlin (1980).

[95] C. M. C. Gambi, M. Carla, D. Senatra – “Comments on a 'Self-Assembly in Fluorocarbon Surfactant Systems”, *Current Opinion in Colloid and Interface Science* **4** (1), 88 – 89 (1999).

[96] D. Makovec, A. Košak, M. Drofenik – “The Preparation of MnZn-Ferrite Nanoparticles in Water-CTAB-Hexanol Microemulsion”, *Nanotechnology* **15**, S160 – 166 (2004).

[97] V. Uskoković, M. Drofenik – “Synthesis of Nanocrystalline Nickel-Zinc Ferrites via a Microemulsion Route”, *Materials Science Forum* **453** – 454, 225 – 230 (2004).

[98] V. Uskoković, M. Drofenik, I. Ban – “The Characterization of Nanosized Nickel-Zinc Ferrites Synthesized within Reverse Micelles of CTAB/1-Hexanol/Water Microemulsion”, *Journal of Magnetism and Magnetic Materials* **284**, 294 – 302 (2004).

[99] J. V. Barth, G. Costantini, K. Kern – “Engineering Atomic and Molecular Nanostructures at Surfaces”, *Nature* **437**, 671 – 679 (2005).

[100] K. Holmberg – “Surfactant-Templated Nanomaterials Synthesis”, *Journal of Colloid and Interface Science* **274**, 355 – 364 (2004).

[101] Filankembo, S. Giorgio, I. Lisiecki, M. P. Pileni – “Is the Anion the Major Parameter in the Shape Control of Nanocrystals”, *Journal of Physical Chemistry B* **107** (30), 7492 – 7500 (2003).

[102] E. Leontidis – “Hofmeister Anion Effects on Surfactant Self-Assembly and the Formation of Mesoporous Solids”, *Current Opinion in Colloid and Interface Science* **7**, 81 – 91 (2002).

[103] W. Kunz, P. Lo Nostro, B. W. Ninham – “The Present State of Affairs with Hofmeister Effects”, *Current Opinion in Colloid and Interface Science* **9**, 1 – 18 (2004).

[104] S. A. Edwards, D. R. M. Williams – “Hofmeister Effects in Colloid Science and Biology Explained by Dispersion Forces: Analytic Results for the

Double Layer Interaction", *Current Opinion in Colloid and Interface Science* **9**, 139 – 144 (2004).

[105] S. Santra, R. Tapec, N. Theodoropoulou, J. Dobson, A. Hebard, W. Tan – "Synthesis and Characterization of Silica-Coated Iron Oxide Nanoparticles in Microemulsion: The Effect of Nonionic Surfactants", *Langmuir* **17**, 2900 – 2906 (2001).

[106] R. I. Nooney, D. Thirunavukkarasu, Y. Chen, R. Josephs, A. E. Ostafin – "Synthesis of Nanoscale Mesoporous Silica Spheres with Controlled Particle Size", *Chemistry of Materials* **14**, 4721 – 4728 (2002).

[107] Y. Qin, K. A. Fichthorn – "Solvation Forces Between Colloidal Nanoparticles: Directed Alignment", *Physical Review E* **73**, 020401 (2006).

[108] J. Yin, Q. Chen, Z. Yin, J. Zhang – "Study on the Oscillation Phenomena of Particle Size Distribution During the Seeded Agglomeration of Sodium Aluminate Liquors", *Light Metals* **173** – 176 (2006).

[109] L. Suber, I. Sondi, E. Matijević, D. V. Goia – "Preparation and the Mechanisms of Formation of Silver Particles of Different Morphologies in Homogeneous Solutions", *Journal of Colloid and Interface Science* **288** (2) 489 – 495 (2005).

[110] A. Grzybowski, G. M. Whitesides – "Three-Dimensional Dynamic Self-Assembly of Spinning Magnetic Disks: Vortex Crystals", *Journal of Physical Chemistry B* **106** (6) 1188 (2002).

[111] P. B. Umbanhowar, F. Melo, H. L. Swinney – "Localized Excitations in a Vertically Vibrated Granular Layer", *Nature* **382**, 793 (1996).

[112] J. M. Ottino – "Complex Systems", *AIChE Journal* **49** (2) 292 – 299 (2003).

[113] Frenkel – "Playing Tricks with Designer 'Atoms'", *Science* **296** (5565) 65 – 66 (2002).

[114] K. N. Pham, A. M. Puertas, J. Bergenholtz, S. U. Egelhaaf, A. Moussaid, P. N. Pusey, A. B. Schofield, M. E. Cates, M. Fuchs, W. C. K. Poon – "Multiple Glassy States in a Simple Model System", *Science* **296** (5565) 104 – 106 (2002).

[115] J. M. Ottino – "New Tools, New Outlooks, New Opportunities", *AIChE Journal* **51** (7) 1840 – 1845 (2005).

[116] J. R. Lead, K. J. Wilkinson – "Aquatic Colloids and Nanoparticles: Current Knowledge and Future Trends", *Environmental Chemistry* **3**, 159 – 171 (2006).

[117] S. D. Škapin, E. Matijević – "Preparation and Coating of Finely Dispersed Drugs 4. Loratadine and Danazol", *Journal of Colloid and Interface Science* **272**, 90 – 98 (2004).

[118] Y. S. Her, E. Matijević, M. C. Chon – "Preparation of Well-Defined Colloidal Barium Titanate Crystals by the Controlled Double-Jet Precipitation", *Journal of Materials Research* **10** (12) 3106 – 3114 (1995).
[119] Goia, E. Matijević – "Precipitation of Barium and Calcium naproxenate Particles of Different Morphologies", *Journal of Colloid and Interface Science* **206** (2) 583 – 591 (1998).
[120] Ruch, E. Matijević – "Preparation of Micrometer Size Budesonide Particles by Precipitation", *Journal of Colloid and Interface Science* **229**, 207 – 211 (2000).
[121] L. Joguet, I. Sondi, E. Matijević – "Preparation of Nanosized Drug Particles by the Coating of Inorganic Cores: Naproxen and Ketoprofen on Alumina", *Journal of Colloid and Interface Science* **251**, 284 – 287 (2002).
[122] J. Wilsdon – "The Politics of Small Things: Nanotechnology, Risk, and Uncertainty", *IEEE Technology and Society Magazine* 16 – 21 (Winter 2004).
[123] ETC Group – "A Tiny Primer on Nano-Scale Technologies and 'The Little Bang Theory'", Action Group on Erosion, Technology and Concentration, Winnipeg, Canada (June 2005); www.etcgroup.org.
[124] Wakai – "Grain Boundary Dynamics in Nanograin Superplasticity, Grain Growth and Sintering", *Materials Science and Technology (MS&T) 2006: Materials and Systems* **2**, 615 – 624 (2006).
[125] S. Edelstein, J. S. Murday, B. B. Rath – "Challenges in Nanomaterials Design", *Progress in Materials Scence* **42**, 5 – 21 (1997).
[126] C. Koch, R. O. Scattergood, K. L. Murty – "Mechanical Behavior of Multiphase Nanocrystalline Materials", *Materials Science and Technology (MS&T) 2006: Materials and Systems* **2**, 347 – 356 (2006).
[127] Y. Yin, P. Alivisatos – "Colloidal Nanocrystal Synthesis and the Organic-Inorganic Interface", *Nature* **437**, 664 – 670 (2005).
[128] R. D. Shull – "Nanomagnetism: A New Materials Frontier", *Materials Science and Technology* (MS&T) 2006 Conference, Cincinnati, OH (October 17, 2006).
[129] R. LeSar – "Is Computational Materials Science Overrated?", Materials Science and Technology (MS&T) 2006 Conference, Cincinnati, OH (October 17, 2006).
[130] V. Uskoković, M. Drofenik – "A Mechanism for the Formation of Nanostructured NiZn Ferrites via a Microemulsion-Assisted Precipitation Method", *Colloids and Surfaces A: Physicochemical and Engineering Aspects* **266**, 168 – 174 (2005).

[131] V. Usković, M. Drofenik – "Mechanism of a Solid-State Formation of $La_{1-x}Sr_xMnO_{3+\delta}$ (0 < x < 0.5) and Magnetic Characterization Thereof", *Materials Science Forum* **518**, 119 – 124 (2006).

[132] V. Usković, M. Drofenik – "Four Novel Co-Precipitation Procedures for the Synthesis of Lanthanum-Strontium Manganites", *Materials & Design* **28** (2) 667 – 672 (2007).

[133] V. Usković, M. Drofenik – "Synthesis of Lanthanum-Strontium Manganites by Oxalate-Precursor Co-Precipitation Methods in Solution and in Reverse Micellar Microemulsion", *Journal of Magnetism and Magnetic Materials* **303** (1) 214 – 220 (2006).

[134] V. Usković, D. Makovec, M. Drofenik – "Synthesis of Lanthanum-Strontium Manganites by a Hydroxide-Precursor Co-Precipitation Method in Solution and Reverse Micellar Microemulsion", *Materials Science Forum* **494**, 155 – 160 (2005).

[135] C. Krishna, G. Venkataiah, S. Ram, P. V. Reddy – "Synthesis and Characterization of Nano Size $La_{0.67}Ca_{0.33}MnO_3$ CMR Material by PVA Chemical Route", *Materials Science and Technology (MS&T) 2006: Materials and Systems* **2**, 507 – 518 (2006).

[136] W. D. Callister Jr. – "Materials Science and Engineering: An Introduction", Fifth Edition, John Wiley & Sons, New York (2000).

[137] V. Usković, A. Košak, M. Drofenik – "Preparation of Silica-Coated Lanthanum-Strontium Manganite Particles with Designable Curie Point, for Application in Hyperthermia Treatments", *International Journal of Applied Ceramic Technology* **3** (2) 134 – 143 (2006).

[138] V. Usković, A. Košak, M. Drofenik – "Silica-Coated Lanthanum-Strontium Manganites for Hyperthermia Treatments", *Materials Letters* **60** (21-22) 2620 – 2622 (2006).

[139] J. C. Lin, J. T. Dipre, M. Z. Yates – "Microemulsion-Directed Synthesis of Molecular Sieve Fibers", *Chemistry of Materials* **15**, 2764 – 2773 (2003).

[140] Debuigne, L. Jeunieau, M. Wiame, J. B. Nagy – "Synthesis of Organic Nanoparticles in Different W/O Microemulsions", *Langmuir* **16**, 7605 – 7611 (2000).

[141] Liu, B. Zuo, A. J. Rondinone, Z. J. Zhang – "Reverse Micelle Synthesis and Characterization of Superparamagnetic $MnFe_2O_4$ Spinel Ferrite Nanocrystallites", *Journal of Physical Chemistry* **104** (6) 1141 - 1145 (2000).

[142] Matijević – "Colloid Science of Ceramic Powders", *Pure & Applied Chemistry* **60** (10) 1479 – 1491 (1988).

[143] C. E. Bunker, B. A. Harruff, P. Pathak, A. Payzant, L. F. Allard, Y.P. Sun – "Formation of Cadmium Sulfide Nanoparticles in Reverse Micelles: Extreme Sensitivity to Preparation Procedure", *Langmuir* **20**, 5642 – 5644 (2004).

[144] Soten, G. A. Ozin – "New Directions in Self-Assembly: Materials Synthesis Over 'All' Length Scales", *Current Opinion in Colloid & Interface Science* **4**, 325 – 337 (1999).

[145] M. Antonietti – "Surfactants for Novel Templating Applications", *Current Opinion in Colloid and Interface Science* **6**, 244 – 248 (2001).

[146] Lisiecki – "Size Control of Spherical Metallic Nanocrystals", *Colloids and Surfaces A: Physicochemical and Engineering Aspects* **250** (1-3) 499 – 507 (2004).

[147] C. E. Fowler, M. Li, S. Mann, H. C. Margolis – "Influence of Surfactant Assembly on the Formation of Calcium Phosphate Materials – A Model for Dental Enamel Formation", *Journal of Materials Chemistry* **15** (32) 3317 – 3325 (2005).

[148] M. Boutonnet, J. Kizling, P. Stenius – "The Preparation of Monodisperse Colloidal Metal Particles from Micro-Emulsions", *Colloids and Surfaces* **5** (3), 209 – 25 (1982).

[149] M. A. Lopez-Quintela – "Synthesis of Nanomaterials in Microemulsions: Formation Mechanisms and Growth Control", *Current Opinion in Colloid & Interface Science* **8**, 137 – 44 (2003).

[150] Kurihara – "Nanostructuring of Liquids at Solid-Liquid Interfaces", *Progress in Colloid and Polymer Science* **121**, 49 – 56 (2002).

[151] Q. Zhong, D. A. Steinhurst, E. E. Carpenter, J. C. Owrutsky – "Fourier Transform Infrared Spectroscopy of Azide Ion in Reverse Micelles", *Langmuir* **18**, 7401 – 7408 (2002).

[152] C. Linehan, J. L. Fulton, R. M. Bean – "Process of Forming Compounds Using Reverse Micelle or Reverse Microemulsion Systems", *US Patent* **5**,770,172 (1998).

[153] R. Zana, J. Lang – "Dynamics of Microemulsions", in *Microemulsions: Structure and Dynamics*, edited by S. E. Friberg and P. Bothorel, Boca Raton: CRC Press, p. 153 – 172 (1987).

[154] Y. Xia, D. Qin, Y. Yin – "Surface Patterning and its Application in Wetting/Dewetting Studies", *Current Opinion in Colloid & Interface Science* **6**, 54 – 56 (2001).

[155] J. C. Warner, A. S. Cannon, K. M. Dye – "Green Chemistry", *Environmental Impact Assessment Review* **24**, 775 – 799 (2004).

[156] D. Anderson, J. L. Anthony, A. Chanda, G. Denison, M. Drolet, D. Fort, M. Joselevich, J. R. Whitfield – "A New Horizon for Future Scientists: Students Voices from the Pan-American Advanced Studies Institute on Green Chemistry", *Green Chemistry* **6**, G5 - 9 (2004).

[157] J. H. Clark – "Green Chemistry: Today (and Tomorrow)", *Green Chemistry* **8**, 17 – 21 (2006).

[158] J. O. Metzger – "Agenda 21 as a Guide for Green Chemistry Research and a Sustainable Future", *Green Chemistry* **6**, G15 - 16 (2004).

[159] Holmberg – "Natural Surfactants", *Current Opinion in Colloid and Interface Science* **6**, 148 – 159 (2001).

[160] S. Lang – "Biological Amphiphiles (Microbial Biosurfactants)", *Current Opinion in Colloid & Interface Science* **7**, 12 – 20 (2002).

[161] Mayer – "Rigid Biological Systems as Models for Synthetic Composites", *Science* **310** (5751) 1144 – 1147 (2005).

[162] H. Pollack – "The Role of Aqueous Interfaces in the Cell", *Advances in Colloid and Interface Science* **103**, 173 – 196 (2003).

[163] P. J. Hiett – "The Place of Life in Our Theories", *BioSystems* 47, 157 – 176 (1998).

[164] K. Richardson, P. Cilliers – "What is Complexity Science? A View from Different Directions", *Emergence* **3** (1) 5 –23 (2001).

[165] J. F. Jenck, F. Agterberg, M. J. Droescher – "Products and Processes for a Sustainable Chemical Industry: A Review of Achievements and Prospects", *Green Chemistry* **6**, 544 – 56 (2004).

[166] K. Kelly – "Out of Control: The New Biology of Machines, Social Systems and the Economic World", Perseus Books, Reading, MA (1994).

[167] P. M. Gleiss, P. F. Stadler, A. Wagner, D. A. Fell – "Relevant Cycles in Chemical Reaction Networks", *Advanced Complex Systems* **1**, 1 – 18 (2001).

[168] Stever – "Science and Technology – Shifting Priorities", *Science, Technology and Modern Society*, edited by F. R. Eirich, Polytechnic Press, New York (1977).

[169] C. Viney – "Self-Assembly as a Route to Fibrous Materials: Concepts, Opportunities and Challenges", *Current Opinion in Solid State and Materials Science* **8**, 95 – 101 (2004).

[170] P. Coussot, Q. D. Nguyen, H. Huynh, D. Bonn – "Viscosity Bifurcation in Thixotropic, Yielding Fluids", *Journal of Rheology* **46** (3) 573 – 589 (2002).

[171] M. Romesin – "Autopoiesis, Structural Coupling and Cognition: A History of These and Other Notions in the Biology of Cognition", *Cybernetics and Human Knowing* **9** (3-4) 5 – 34 (2002).

[172] P. A. Bachmann, P. Walde, P. L. Luisi, J. Lang – "Self-Replicating Micelles - Aqueous Micelles and Enzymatically Driven Reactions in Reverse Micelles", *Journal of the American Chemical Society* **113** (22) 8204 – 8209 (1991).

[173] P. A. Bachmann, P. L. Luisi, J. Lang – "Self-Replicating Reverse Micelles", *Chimia* **45** (9) 266 – 268 (1991).

[174] P. A. Bachmann, P. L. Luisi, J. Lang – "Autocatalytic Self-Replicating Micelles as Models for Prebiotic Structures", *Nature* **357**, 57 – 59 (1992).

[175] Tuck – "The Role of Atmospheric Aerosols in the Origin of Life", *Surveys in Geophysics* **23**, 379 – 409 (2002).

[176] Carlile, G. D. Rees, B. H. Robinson, T. D. Steer, M. Svensson – "Lipase-Catalysed Interfacial Reactions in Reverse Micellar Systems - Role of Water and Microenvironment in Determining Enzyme Activity or Dormancy", *Journal of the Chemical Society – Faradey Transactions* **92** (23) 4701-4708 (1996).

[177] Y. X. Chen, X. Z. Zhang, K. Zheng, S. M. Chen, Q. C. Wang, X. X. Wu – "Protease-Catalyzed Synthesis of Precursor Dipeptides of RGD with Reverse Micelles", *Enzyme and Microbial Technology* **23** (3-4): 243-248 (1998).

[178] P. P. Sun, Y. M. Zhang – "The Catalysis and Asymmetric Induction of Chiral Reverse Micelle: Synthesis of Optically Active Alpha-Amino Acids", *Synthetic Communications* **27** (23): 4173-4179 (1997).

[179] K. B. Lee, L. H. Poppenborg, D. C. Stuckey – "Terpene Ester Production in a Solvent Phase Using a Reverse Micelle-Encapsulated Lipase", *Enzyme and Microbial Technology* **23** (3-4): 253-260 (1998).

[180] R. Babu, P. F. Flynn, A. J. Wand – "Preparation, Characterization, and NMR Spectroscopy of Encapsulated Proteins Dissolved in Low Viscosity Fluids", *Journal of Biomolecular NMR* **25** (4): 313-323 (2003).

[181] E. P. Melo, M. R. Aires-Barros, J. M. Cabral – "Reverse Micelles and Protein Biotechnology", *Biotechnology Annual Review* **7,** 87 – 129 (2001).

[182] E. P. Melo, S. M. B. Costa, J. M. S. Cabral, P. Fojan, S. B. Petersen – "Cutinase – SOT Interactions in Reverse Micelles: The Effect of 1-Hexanol", *Chemistry and Physics of Lipids* **124**, 37 – 47 (2003).

[183] G. Chang, T. M. Huang, H. C. Hung – "Reverse Micelles as Life-Mimicking Systems", *Proceedings of the National Scientific Council ROC(B)* **24** (3) 89 - 100 (1999).

[184] P. L. Luisi, P. Walde, T. Oberholzer – "Lipid Vesicles as Possible Intermediates in the Origin of Life", *Current Opinion in Colloid & Interface Science* **4**, 33 – 39 (1999).

[185] P. Baglioni, D. Berti – "Self Assembly in Micelles Combining Stacking and H-Bonding", *Current Opinion in Colloid and Interface Science* **8**, 55 – 61 (2003).

[186] W. M. Tolles – "Self-Assembled Materials", *MRS Bulletin* **25** (10) 36 – 38 (2000).

[187] "Molecular Self-Assembly: An Interview with Professor Gero Decher", *Essential Science Indicators Special Topics*, retrieved from http://www.esi-topics.com/msa/interviews/GeroDecher.html (September 2002).

[188] Brancato, F. Coutrot, D. A. Leigh, A. Murphy, J. K. Y. Wong, F. Zerbetto – "From Reactants to Products via Simple Hydrogen-Bonding Networks: Information Transmission in Chemical Reactions", *Proceedings of the National Academy of Sciences of the United States of America* **99** (8) 4967 – 4971 (2002).

[189] Tirrell – "Modular Materials by Self-Assembly", *AIChE Journal* **51** (9) 2386 – 2390 (2005).

[190] A. Leser – "Food Colloids: Learning from Nature", *Current Opinion in Colloid and Interface Science* **11**, 133 – 134 (2006).

[191] E. Dickinson – "Food Colloids... Drifting into the Age of Nanoscience", *Current Opinion in Colloid and Interface Science* **8**, 346 – 348 (2003).

[192] D. Velev, A. M. Lenhoff – "Colloidal Crystals as Templates for Porous Materials", *Current Opinion in Colloid and Interface Science* **5**, 56 – 63 (2000).

[193] A. Campbell – "Interfacial Regulation of Crystallization in Aqueous Environments", *Current Opinion in Colloid & Interface Science* **4**, 40 – 45 (1999).

[194] Maturana, F. Varela – "The Tree of Knowledge: The Biological Roots of Human Understanding", Shambhala, Boston, MA (1998).

[195] R. Villegas, A. Redjaimia, M. Confente, M. T. Perrot-Simonetta – "On the Fractal Nature of Acicular Ferrite, and V(C,N) Precipitation in Medium Carbon Microalloyed Steels", *Materials Science and Technology (MS&T) 2006: Materials and Systems* **2**, 401 – 410 (2006).

[196] G. Decher – "Fuzzy Nanoassemblies: Toward Layered Polymeric Multicomposites", *Science* **277** (5330), 1232 – 1237 (1997).

[197] Y. Dauphin – "Structures, Organo-Mineral Compositions and Diagenetic Changes in Biominerals", *Current Opinion in Colloid & Interface Science* **7**, 133 – 138 (2002).

[198] Kirkham, S. J. Brookes, R. C. Shore, S. R. Wood, D. A. Smith, J. Zhang, H. Chen, C. Robinson – "Physico-Chemical Properties of Crystal Surfaces in Matrix-Mineral Interactions During Mammalian Biomineralisation", *Current Opinion in Colloid & Interface Science* **7**, 124 – 132 (2002).
[199] "Editorial: Not-So-Deep Impact", *Nature* **435**, 1003 – 1004 (2005).
[200] Schickore – "'Through Thousands of Errors We Reach the Truth' – But How? On the Epistemic Roles of Error in Scientific Practice", *Studies in History and Philosophy of Modern Physics* **36**, 539 – 56 (2005).
[201] "Editorial: Wiki's Wild World", *Nature* **438** (7070) 890 (2005).
[202] Rueda, E. Cubero, C. A. Laughton, M. Orozco – "Exploring the Counterion Atmosphere around DNA: What Can Be Learned from Molecular Dynamics Simulations?", *Biophysical Journal* **87** (2) 800 – 811 (2004).
[203] A. Kornyshev, A. Wynveen – "Nonlinear Effects in the Torsional Adjustment of Interacting DNA", *Physical Review E* **69**, 041905 (2004).
[204] Peyrard – "Nonlinear Dynamics and Statistical Physics of DNA", *Nonlinearity* **17**, R1 – R40 (2004).
[205] M. Dimiduk, C. Woodward, R. LeSar, M. D. Uchic – "Scale-Free Intermittent Flow in Crystal Plasticity", *Science* **312** (5777) 1188 – 1190 (2006).
[206] Gleick – "Chaos: Making a New Science", Heinemann, London (1987).
[207] Q. Peng, J. G. Wu, R. D. Soloway, T. D. Hu, W. D. Huang, Y. Z. Xu, L. B. Wang, X. F. Li, W. H. Li, D. F. Xu, G. X. Xu – "Periodic and Chaotic Precipitation Phenomena in Bile Salt System Related to Gallstone Formation", *Biospectroscopy* **3** (3) 195 – 205 (1997).
[208] T. Yano – "Fractal Nature of Food Materials", *Bioscience, Biotechnology, and Biochemistry* **60** (5) 739 – 744 (1996).
[209] R. Rajagopalan – "Simulations of Self-Assembling Systems", *Current Opinion in Colloid and Interface Science* **6**, 357 – 365 (2001).
[210] T. M. Bayerl – "Collective Membrane Motions", *Current Opinion in Colloid and Interface Science* **5**, 232 – 236 (2000).
[211] Sain, M. Wortis – "The Influence of Tether Dynamics on Forced Kramers Escape from a Kinetic Trap", *Physical Review E* **70**, 031102 (2004).
[212] G. Margolin, E. Barkai – "Single-Molecule Chemical reactions: Reexamination of the Kramers Approach", *Physical Review E* **72**, 025101 (2005).
[213] Frenkel – "Soft Condensed Matter", *Physica A* **313**, 1 – 3 (2002).
[214] C. Shelley, M. Y. Shelley – "Computer Simulation of Surfactant Solutions", *Current Opinion in Colloid and Interface Science* **5**, 101 – 110 (2000).

[215] Dugić, D. Raković, M. Plavšić – "The Polymer Conformational Stability and Transitions: A Quantum Decoherence Theory Approach", *Surfactant Science Series: Finely Dispersed Particles* **130**, 217 – 231 (2006).

[216] Raković, M. Dugić, M. Plavšić, G. Keković, I. Ćosić, D. Davidović – "Quantum Decoherence and Quantum-Holographic Information Processes: From Biomolecules to Biosystems", *Materials Science Forum* **518**, 485 – 490 (2006).

[217] McFadden – "Quantum Mechanics and Biology", in *Quantum: A Guide for the Perplexed*, J. Al-Khalili (ed.), Weidenfeld & Nicolson, London (2003).

[218] R. Ettelaie – "Computer Simulation and Modeling of Food Colloids", *Current Opinion in Colloid and Interface Science* **8**, 415 – 421 (2003).

[219] Capra – "The Tao of Physics: An Exploration of the Parallels between Modern Physics and Eastern Mysticism", Shambhala, Boston, MA (1983).

[220] D. Bohm – "Wholeness and the Implicate Order", Ark Paperbacks, London (1980).

[221] E. Laszlo – "Nonlocal Coherence in the Living World", *Ecological Complexity* **1**, 7 – 15 (2004).

[222] Capra – "The Hidden Connections: Integrating the Biological, Cognitive and Social, Dimensions of Life into a Science of Sustainability", Doubleday, New York (2002).

[223] J. Wu, D. Bratko, J. M. Prausnitz – "Interaction Between Like-Charged Colloidal Spheres in Electrolyte Solutions", *Proceedings of the National Academy of Sciences of the United States of America* **95** (26) 15169 – 15172 (1998).

[224] Z. Y. Zhang, D. C. Langreth, J. P. Perdew – "Planar-Surface Charge Densities and Energies Beyond the Local-Density Approximation", *Physical Review B* **41**, 5674 – 5684 (1990).

[225] Schmitz – "Fluctuations in a Nonequilibrium Colloidal Suspensions", *Physica A* **206** (1-2) 25 – 57 (1994).

[226] E. Levinger – "Water in Confinement", *Science* **298**, 1722 – 1723 (2002).

[227] D. M. Engelman – "Membranes are More Mosaic than Fluid", *Nature* **438**, 578 – 580 (2005).

[228] S. T. Hyde, G. E. Schröder – "Novel Surfactant Mesostructural Topologies: Between Lamellae and Columnar (Hexagonal) Forms", *Current Opinion in Colloid and Interface Science* **8**, 5 – 14 (2003).

[229] R. Piazza – "Cradles for Life: Self-Organization of Biological Colloids", *Current Opinion in Colloid and Interface Science* **11** (1), 30 – 34 (2006).

[230] N. Levine – "Physical Chemistry", McGraw-Hill, New York (1978).

[231] E. von Glasersfeld – "Radical Constructivism: A Way of Knowing and Learning", RoutledgeFalmer, London (1995).
[232] Frenot, I. S. Chronakis – "Polymer Nanofibers Assembled by Electrospinning", *Current Opinion in Colloid and Interface Science* **8**, 64 – 75 (2003).
[233] S. D. Tzeng, K. J. Lin, J. C. Hu, L. J. Chen, S. Gwo – "Templated Self-Assembly of Colloidal Nanoparticles Controlled by Electrostatic Nanopatterning on $Si_3N_4/SiO_2/Si$ Electret", *Advanced Materials* **18** (9) 1147 – 1151 (2006).
[234] E. Bourgeat–Lami – "Organic-Inorganic Nanostructured Colloids", *Journal of Nanoscience and Nanotechnology* **2** (1) 1 – 24 (2002).
[235] M. Mendes, J. A. Preece – "Precision Chemical Engineering: Integrating Nanolithography and Nanoassembly", *Current Opinion in Colloid and Interface Science* **9**, 236 – 248 (2004).
[236] G. M. Whitesides – "The Origins and the Future of Microfluidics", *Nature* **442** (7101) 368 – 373 (2006).
[237] K. Arora, B. V. R. Tata – "Interactions, Structural Ordering and Phase Transitions in Colloidal Dispersions", *Advances in Colloid and Interface Science* **78**, 49 – 97 (1998).
[238] W. T. Shin, S. Yiacoumi, C. Tsouris – "Electric-Field Effect on Interfaces: Electrospray and Electrocoalescence", *Current Opinion in Colloid and Interface Science* **9**, 249 – 255 (2004).
[239] S. Krishnamoorthy, C. Hinderling, H. Heinzelmann – "Nanoscale Patterning with Block Copolymers", *Materials Today* **9** (9) 40 – 47 (2006).
[240] Li, C. K. Ober – "Block Copolymers: Patterns and Templates", *Materials Today* **9** (9) 30 – 39 (2006).
[241] P. Stoykovich, P. F. Nealey – "Block Copolymers and Conventional Lithography", *Materials Today* **9** (9) 20 – 29 (2006).
[242] R. Glass, M. Möller, J. P. Spatz – "Block Copolymer Micelle Nanolithography", *Nanotechnology* **14**, 1153 – 1160 (2003).
[243] Gambardella, M. Blanc, H. Brune, K. Kuhnke, K. Kern – "One-Dimensional Metal Chains on Pt Vicinal Surfaces", *Physical Review B* **61**, 2254 – 2262 (2000).
[244] E. F. Schumacher – "Small is Beautiful: Economics as if People Mattered. 25 Years Later...with Commentaries", Hartley & Marks, Vancouver (1998).
[245] T. Winograd, F. Flores – "Understanding Computers and Cognition: A New Foundations for Design", Ablex Publishing Corporation, Norwood, NJ (1987).

INDEX

D

E

F

G

H

I

K

L

M

N

O

P

Q

R

S

T

U

V

W

Y

Z